Nanostructured Materials
and Nanotechnology

Nanostructured Materials and Nanotechnology

*A Collection of Papers Presented at the
31st International Conference on Advanced
Ceramics and Composites
January 21–26, 2007
Daytona Beach, Florida*

Editors
Sanjay Mathur
Mrityunjay Singh

Volume Editors
Jonathan Salem
Dongming Zhu

The
American
Ceramic
Society

WILEY-INTERSCIENCE
A John Wiley & Sons, Inc., Publication

Published by John Wiley & Sons, Inc., Hoboken, New Jersey.
Published simultaneously in Canada.

For general information on our other products and services or for technical support, please contact our Customer Care Department within the United States at (800) 762-2974, outside the United States at (317) 572-3993 or fax (317) 572-4002.

Wiley also publishes its books in a variety of electronic formats. Some content that appears in print may not be available in electronic format. For information about Wiley products, visit our web site at www.wiley.com.

Wiley Bicentennial Logo: Richard J. Pacifico

Library of Congress Cataloging-in-Publication Data is available.

ISBN 978-0-470-19637-3

10 9 8 7 6 5 4 3 2 1

Contents

Preface

The First International Symposium on Nanostructured Materials and Nanotechnology was held during the 31st International Conference on Advanced Ceramics and Composites, in Daytona Beach, FL, January 21–26, 2007.

The major motivation behind this effort was to create an international platform within ICCAC focusing on science, engineering and manufacturing aspects in the area of nanostructured materials. The symposium covered a broad perspective including synthesis, processing, modeling and structure-property correlations in nanomaterials. More than 110 contributions including 22 invited talks, 68 oral presentations and 21 posters, were presented from 17 countries (USA, Japan, Germany, Canada, Korea, India, Ireland, Turkey, Slovenia, Sweden, Italy, Spain, New Zealand, Belgium, Thailand, Singapore and the United Kingdom). The speakers represented universities, research institutions, and industry. In order to make this symposium an attractive forum for interdisciplinary presentations and discussions, it was decided to have a focused session on industrial development and applications of nanomaterials, where invited speakers from global leaders such as BASF, Bayer and Buehler Limited presented the industrial outlook and future perspectives.

This issue contains peer-reviewed (invited and contributed) papers incorporating latest developments related to synthesis, processing and manufacturing technologies of nanoscaled materials including nanoparticle-based composites, electrospinning of nanofibers, functional thin films, ceramic membranes and self-assembled functional nanostructures and devices. These papers discuss several important aspects related to fabrication and engineering issues necessary for understanding and further development of processing and manufacturing of nanostructured materials and systems.

The editors wish to extend their gratitude and appreciation to all the authors for their cooperation and contributions, to all the participants and session chairs for their time and efforts, and to all the reviewers for their valuable comments and suggestions. Financial support from ItN Nanovation, Ltd., Saarbruecken, Germany as well as the Engineering Ceramic Division of the American Ceramic Society is gratefully acknowledged. Thanks are due to the staff of the meetings and publication departments of The American Ceramic Society for their invaluable assistance.

We hope that this issue will serve as a useful reference for the researchers and technologists working in the field of interested in science and technology of nanostructured materials and devices.

SANJAY MATHUR
Leibniz Institute of New Materials
Saarbruecken, Germany

MRITYUNJAY SINGH
Ohio Aerospace Institute
Cleveland, Ohio, USA

Introduction

2007 represented another year of growth for the International Conference on Advanced Ceramics and Composites, held in Daytona Beach, Florida on January 21-26, 2007 and organized by the Engineering Ceramics Division (ECD) in conjunction with the Electronics Division (ED) of The American Ceramic Society (ACerS). This continued growth clearly demonstrates the meetings leadership role as a forum for dissemination and collaboration regarding ceramic materials. 2007 was also the first year that the meeting venue changed from Cocoa Beach, where it was originally held in 1977, to Daytona Beach so that more attendees and exhibitors could be accommodated. Although the thought of changing the venue created considerable angst for many regular attendees, the change was a great success with 1252 attendees from 42 countries. The leadership role in the venue change was played by Edgar Lara-Curzio and the ECD's Executive Committee, and the membership is indebted for their effort in establishing an excellent venue.

The 31st International Conference on Advanced Ceramics and Composites meeting hosted 740 presentations on topics ranging from ceramic nanomaterials to structural reliability of ceramic components, demonstrating the linkage between materials science developments at the atomic level and macro level structural applications. The conference was organized into the following symposia and focused sessions:

- Processing, Properties and Performance of Engineering Ceramics and Composites
- Advanced Ceramic Coatings for Structural, Environmental and Functional Applications
- Solid Oxide Fuel Cells (SOFC): Materials, Science and Technology
- Ceramic Armor
- Bioceramics and Biocomposites
- Thermoelectric Materials for Power Conversion Applications
- Nanostructured Materials and Nanotechnology: Development and Applications
- Advanced Processing and Manufacturing Technologies for Structural and Multifunctional Materials and Systems (APMT)

- Porous Ceramics: Novel Developments and Applications
- Advanced Dielectric, Piezoelectric and Ferroelectric Materials
- Transparent Electronic Ceramics
- Electroceramic Materials for Sensors
- Geopolymers

The papers that were submitted and accepted from the meeting after a peer review process were organized into 8 issues of the 2007 Ceramic Engineering & Science Proceedings (CESP); Volume 28, Issues 2-9, 2007 as outlined below:

- Mechanical Properties and Performance of Engineering Ceramics and Composites III, CESP Volume 28, Issue 2
- Advanced Ceramic Coatings and Interfaces II, CESP, Volume 28, Issue 3
- Advances in Solid Oxide Fuel Cells III, CESP, Volume 28, Issue 4
- Advances in Ceramic Armor III, CESP, Volume 28, Issue 5
- Nanostructured Materials and Nanotechnology, CESP, Volume 28, Issue 6
- Advanced Processing and Manufacturing Technologies for Structural and Multifunctional Materials, CESP, Volume 28, Issue 7
- Advances in Electronic Ceramics, CESP, Volume 28, Issue 8
- Developments in Porous, Biological and Geopolymer Ceramics, CESP, Volume 28, Issue 9

The organization of the Daytona Beach meeting and the publication of these proceedings were possible thanks to the professional staff of The American Ceramic Society and the tireless dedication of many Engineering Ceramics Division and Electronics Division members. We would especially like to express our sincere thanks to the symposia organizers, session chairs, presenters and conference attendees, for their efforts and enthusiastic participation in the vibrant and cutting-edge conference.

ACerS and the ECD invite you to attend the 32nd International Conference on Advanced Ceramics and Composites (http://www.ceramics.org/meetings/daytona2008) January 27 - February 1, 2008 in Daytona Beach, Florida.

JONATHAN SALEM AND DONGMING ZHU, Volume Editors
NASA Glenn Research Center
Cleveland, Ohio

FRACTIONATION OF NANOCRYSTALLINE TiO_2 BY COAGULATION OF HYDROSOLS

Pavlova-Verevkina O.B., Ozerina L.A., Chvalun S.N.
Karpov Institute of Physical Chemistry
Vorontsovo Pole St. 10
Moscow, Russia, 105064

Ozerin A.N.
Institute of Synthetic Polymeric Materials
Profsoyuznaya St. 70
Moscow, Russia, 117393

ABSTRACT
    From the polydisperse TiO_2 hydrosols contained plate-like anatase nanocrystals the narrow nanocrystal fractions have been isolated. The anatase platelets in the fractions have been shown to have similar thickness ~2 nm and different lateral sizes in the range of 4-29 nm. The gels and xerogels ordered in nanometer scale have been prepared from the isolated fractions. The structure of gels has been shown to depend crucially on acid concentration in the dispersion medium. Destabilization of diluted TiO_2 sol by some electrolytes and the slow increase of sol turbidity have been studied. It has been found that the rate of the slow turbidity increase rises strongly with electrolyte concentration and has a minimum at some pH-values. The results obtained enable to improve the fractionation procedure and to develop highly ordered TiO_2 nanomaterials.

INTRODUCTION
    Uniform (in size, shape, and phase composition) oxide nanocrystals can be used as building blocks for the preparation of highly ordered materials of nanometer periodicity [1]. These materials may be formed by drying the liquid dispersions of uniform nanocrystals. It is obviously that such dispersions are exclusive objects to study the effects of different parameters on structure and properties. So it is very important to develop methods of preparation of the uniform nanocrystals and their stable dispersions.
    We developed the procedure of fractionation of TiO_2 nanocrystals based on the coagulation of polydisperse hydrosols by strong acids [2, 3]. The fractionation is possible due to the reversibility of the rapid coagulation and to the existence of a strong dependency of a threshold of the rapid coagulation C_c on particle size. To obtain more narrow TiO_2 fractions it is necessary to study the peculiarities of interaction of the sols with electrolytes in more detail.
    In the paper we describe the preparation of narrow TiO_2 fractions of different dispersity from polydisperse hydrosols, the preparation of the ordered concentrated dispersions from the fractions, and also some new experimental results on destabilization of the TiO_2 sols by electrolytes.

EXPERIMENTAL
    Initial polydisperse TiO_2 hydrosols synthesized from Ti tetrabutoxide and tetrachloride were stabilized by HCl and HNO_3 and contained mainly the anatase polymorph nanoparticles [2-4]. A mean hydrodynamic radius of particles R_h in the sols was in the range of 7-10 nm.

R_h was measured by DLS using photon-correlation spectrometer; optical density D and coagulation thresholds C_c were determined by UV-visible spectroscopy in a 30-mm wide cell [2-3]. TiO$_2$ weight concentration in the dispersions was determined by the gravimetric analysis.

The fractionation was conducted by the stepped coagulation using the same acid, which stabilized the initial sol. From the precipitated fractions at first the stable 5-15% sols of different dispersity were prepared. Then, after determination of TiO$_2$ content and pH, the sols were diluted or concentrated [3]. Gels were obtained either by evaporating the concentrated sols in the air or by coagulation of the sols by HCl additives. Xerogels were prepared by drying the sols and gels in the air at room temperature till constant weight.

All stable dilute and concentrated dispersions prepared were studied in some weeks after preparation.

The X-Rays experiments in small and wide angles of scattering are described in detail in [3]. Diffraction patterns of the fractions in wide angles were obtained for dried sols. A mean size of primary anatase nanocrystals L_{101} was calculated from a peak halfwidth by the Scherer equation.

The SAXS measurements of the sols were carried in capillaries, while the gels were investigated in a flat cuvette. Scattering coordinate was measured in terms of the scattering vector modulus $s = 4\pi \sin\theta/\lambda$ in the range of s from 0.07 to 4.26 nm^{-1}. The preliminary treatment of the initial scattering curves (smoothing, normalization) as well as the calculation of the correlation function of the scattering particles and of the pair distribution function within a single particle was made with SYRENA software complex [5]. Particle radius of gyration R_g was calculated from the innermost part of the SAXS curve (Guinier approximation [6]) and from Fourier transformation of the SAXS curve [5], then the values obtained were averaged.

RESULTS AND DISCUSSION

FRACTIONATION

Five anatase TiO$_2$ fractions of different dispersity were obtained by the stepped coagulation of the initial hydrosol by HNO$_3$. The coagulation thresholds C_c of these fractions ranged in an interval 0.7-1.9 M HNO$_3$, the R_h values - from 4 to 12 nm, and the L_{101} values - from 3 to 6 nm (see Table).

Table. The characteristics of five nanocrystalline TiO$_2$ fractions stabilized by HNO$_3$. Samples 1-5 were studied in 1 month after preparation, sample 5a in 2 years.

Fraction	C_c	a	b	c	R_g	R_h	L_{101}
	M HNO$_3$		nm		nm	nm	nm
1	1.6-1.9	15	4.3	1.7	4.3	4	2.9
2	1.3-1.6	16	4.8	1.7	4.9	5	3.9
3	1.0-1.3	21	5.4	1.7	6.2	6	4.1
4	0.8-1.0	24	7.4	2.0	7.2	7	4.5
5	0.7-0.8	28	8.3	2.0	8.4	12	5.6
5a	0.7-0.8	29	9.9	1.8	9.0	14	5.6

Then the five 1% sols stabilized by HNO$_3$ with pH 1 were prepared from the fractions and studied by SAXS. Figure 1 shows the scattering curves. Their analysis allowed to determine the gyration radius R$_g$ and a shape of nanoparticles in the fractions (see Table).

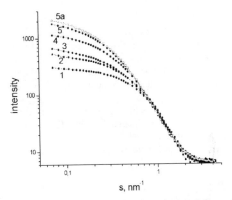

Figure 1. SAXS curves of the five TiO$_2$ fractions isolated from the polydisperse sol. Curves 5 and 5a refer to the same sol measured in 1 month and 2 years correspondingly.

To determine a shape of nanoparticles the correlation functions of the scattering particles $\gamma(r)$ and the pair distance distribution functions within a single scattering particle $f(r) = (\gamma(r) * r)$ and $p(r) = (\gamma(r) * r^2)$ were calculated [5]. For example the $p(r)$ functions are presented in Figure 2. It was established that the pair distribution functions were fitted well with the appropriate functions for the particles with a prism-like shape rather than for the globular-like or rod-like particles. Under the assumption of prism-like particles scattering, the characteristic sizes of a prism **a**, **b** and **c** were calculated.

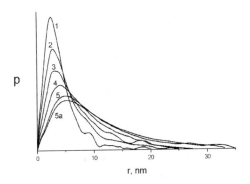

Figure 2. p(R) functions calculated from the curves presented in Figure 1.

The Table shows that in all fractions studied the anatase nanoparticles are plate-like and the platelets have minimum constant size c close to 2 nm. The other two sizes of the platelets, a and b, in different fractions range from 4 to 29 nm. It is also seen that the higher coagulation threshold C_c the lower particles sizes a, b, R_g, R_h, and L_{101}.

ORDERED TIO₂ DISPERSIONS

Gels and solid xerogels stabilized by HCl were prepared from narrow TiO₂ fractions and studied by SAXS. The analysis of the results revealed that some samples have rather ordered structure at nanometer scale. The scattering curves of these samples have a broad maximum which position corresponds to a long period in the range of 6-30 nm. It was established that the character of ordering of the sols and gels depends crucially on HCl concentration C in the dispersion medium.

In Figure 3 the scattering curves of three gels contained uniform plate-like anatase TiO₂ nanocrystals are presented (curves 1-3). These gels have HCl concentration C equal to 0.2, 0.8, and 1.6 M, correspondingly, and relatively close concentration of TiO₂ (57-67 w.%) and R_h values (6-8 nm). We can see that the scattering curves of these dispersions differ qualitatively.

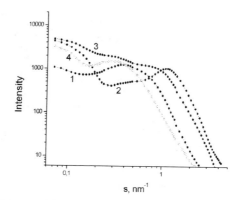

Figure 3. SAXS curves of the gels with HCl concentration C 0.2 (1), 0.8 (2) and 1.6 M (3) and xerogel (4), prepared from narrow anatase TiO₂ fractions.

Curve 4 in Figure 3 corresponds to a xerogel obtained by slow drying one of the gels in the air. This glass-like solid sample contains 85 w.% TiO₂. The distinct maximum on the scattering curve reveals that this material is also relatively ordering.

The development of such approach will open the opportunity to prepare more ordered TiO₂ dispersions and nanomaterials of different structure.

DESTABILIZATION OF THE SOLS BY ELECTROLYTES

Kinetics of destabilization by HCl and KCl of the dilute sols prepared from the narrow fractions was studied as well. It was established that just after the electrolyte has been introduced into the sol the slow monotonic increase of turbidity occurs following by partial sedimentation.

At the first stage of this slow process, which can last for many months, the sols with a total concentration of electrolyte C<C_c remain uniform. In contrast to the rapid coagulation taking place at C>C_c and being entirely reversible, the slow process occurs both at low and high electrolyte concentrations and seems to be partially reversible.

Figure 4 shows the time dependencies of turbidity for five 0.4% sols stabilized by HCl after introducing additives of HCl (curve 1) or KCl (curves 2-5). All sols were prepared from the same narrow fraction characterized by R_h=6 nm. Sols with KCl additives have different pH values equal to 2.2 (curve 2) and 0.9 (curves 3-5); sol with HCl additive has pH~0. It was found that during the experiment the pH values of the samples did not change.

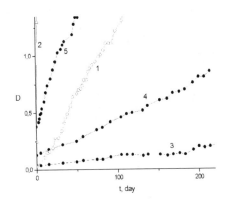

Figure 4. Kinetics of optical density of the 0.4 % sols with different electrolyte concentration and pH. 1 - 1.4 M HCl, pH~0; 2 - 0.8 M KCl, pH 2.2; 3 - 0.8 M KCl, pH 0.9; 4 - 1.1 M KCl, pH 0.9; 5 - 1.4 M KCl, pH 0.9.

From curves 3-5 in Figure 4 one can see that a rate of the turbidity increase rises strongly with electrolyte concentration at pH=const. Comparison of the curves 1-3 shows that the rate depends essentially also on pH: it is minimal in the sols with pH~1 and increases in orders of magnitude at pH ~0 and ~2.

The mechanism of these slow structural changes in the TiO₂ dispersions in presence of electrolytes is not clear yet. To understand it the additional experiments are necessary. The further investigation in this direction is very crucial in improving the fractionation procedure and in manipulation in nanoparticle size and shape.

CONCLUSIONS
1. From the polydisperse hydrosol contained plate-like anatase TiO₂ nanoparticles of 2 nm thick some fractions of platelets different in their lateral sizes have been isolated.
2. The gels and xerogels ordered in nanometer scale have been prepared from the narrow TiO₂ fractions. The structure of the gels has been shown to depend crucially on the acid content in the dispersion medium.

3. Study of the destabilization of the TiO$_2$ sols by HCl and KCl has revealed that the rate of the slow increase of turbidity rises strongly with electrolyte concentration and has a minimum at pH equal to ~1.

REFERENCES

[1] M. Niederberger and H. Cölfen. Oriented attachment and mesocrystals: Non-classical crystallization mechanisms based on nanoparticle assembly. *Phys. Chem. Chem. Phys.*, **8**, 3271–87 (2006).

[2] O.B. Pavlova-Verevkina, Yu.A. Shevchuk and V.V. Nazarov, Coagulation peculiarities and fractionation of nanodispersed titanium dioxide hydrosol. *Colloid Journal*, **65**, No. 4, 474-77 (2003).

[3] O.B. Pavlova-Verevkina, S.N. Chvalun, E.D. Politova, V.V. Nazarov, L.A. Ozerina and A.N. Ozerin, Study of the stable nanocrystalline TiO$_2$ hydrosol and its fractions. *Journal of Sol-Gel Science and Technology*, **35**, 91-97 (2005).

[4] O.B. Pavlova-Verevkina, N.V. Kul'kova, E.D. Politova, Yu.A. Shevchuk and V.V. Nazarov. Preparation of thermostable highly dispersed titanium dioxide from stable hydrosols. *Colloid journal*, **65**, No 2., 226-29 (2003).

[5] L.A. Feigin and D.I. Svergun, *Structure Analysis by Small-Angle X-ray and Neutron Scattering,* Plenum Press, New-York (1987).

[6] A.Guinier and G. Fournet, *Small Angle Scattering of X-rays,* Wiley, New York (1955).

SYNTHESIS OF NANOCRYSTALLINE FORSTERITE (Mg$_2$SiO$_4$) VIA POLYMER MATRIX ROUTE

Ali Saberi
Department of Material Science and Engineering
Faculty of Mechanical Engineering, University of Tabriz
51666-16471 Tabriz, Iran

Zahra Negahdari
Chair of Materials Processing
Faculty of Applied Science, University of Bayreuth
D-95447 Bayreuth, Germany

Babak Alinejhad, Framarz Kazemi, Ali Almasi
Department of Material Science and Engineering
Iran University of Science and Technology (IUST)
16846-13114 Tehran, Iran

ABSTRACT

Nanocrystalline forsterite (Mg$_2$SiO$_4$) powder was synthesized using sucrose as a chelating agent and template material from an aqueous solution of magnesium nitrate and colloidal silica. The synthesized powders were characterized by X-ray diffraction (XRD), fourier transform infrared spectroscopy (FT-IR), simultaneous thermal analysis (STA), and field emission electron microscope (FEM). The synthesised nano-powder had particle size smaller than 200nm and average crystallite size of powders calcined at 800°C for 3h was in the range of 20nm.

INTRODUCTION

Forsterite is a crystalline magnesium silicate with chemical formula Mg$_2$SiO$_4$ which has extremely low electrical conductivity that makes it an ideal substrate material for electronics. On the other hand it shows good refractoriness due to high melting point ($\approx$1890°C), low thermal expansion, good chemical stability and excellent insulation properties even at high temperatures.[1, 2]

Forsterite has been synthesized by different methods such as solid state reaction, self-propagation high-temperature synthesis, and sol–gel. The production of forsterite via solid-state reactions usually requires high temperature and long reaction time while the sol–gel process can provide molecular-level of mixing and high degree of homogeneity, which leads to reduce crystallization temperature and prevent from phase segregation during heating. However, in multi-component silicate systems, the hydrolysis and condensation rates are different with in silica and the other alkoxides which may cause non-uniform precipitation and chemical inhomogeneity of the gels, and also result in higher crystallization temperature and undesired phases.[3-5]

One of the effective methods that recently have been developed is sucrose process, which is established for the preparation of fine oxide ceramic powders.[6] In this technique, a uniform particle size powder is produced due to a homogeneous metal ion distribution in the solution. As mean while, other elements such as C, H, and N are easily removed during calcination. Therefore, the purity of the final powder does not get affected when sucrose is used as a chelating agent and template material.[7] The object of the present research is the synthesis of nanocrystalline forsterite powder from magnesium nitrate, colloidal silica, and sucrose at low temperatures.

EXPERIMENTAL

Magnesium nitrate $(Mg(NO_3)_2 \cdot 6H_2O)$, Sucrose as a template material, PVA (poly vinyl alcohol, MW=145000), Nitric acid (all have been supplied by Merck Co., Germany) and Colloidal Silica with particle size smaller than 14 nm (26wt.% solid fraction-from Monatso Co., Belgium) were used as starting materials.

To prepare a transparent sol, 0.0142 mole magnesium nitrate (3.639gr) was dissolved in 50 ml of deionized water. Then 0.0071 mole of silica (1.642gr of colloidal silica) was introduced into the solution to set MgO/SiO_2 molar ratio to 2. Sucrose solution was prepared separately by adding 0.0568 mole (19.426gr) of it into 100 ml of deionized water. Two solutions were then mixed together and continuously agitated by magnetic hot plate stirer for 2 hr. PVA solution was prepared by mixing of 0.0071 mole (0.312gr) of PVA with 20 ml of deionized water and then was added into the final solution and pH was adjusted to 1 by drop wise addition of diluted nitric acid and mixing continued for 4 hr. The mole ratio of the Mg^{2+} ions, sucrose and PVA monomer was 1:4:0.5.

Nitric acid addition breaks sucrose into glucose and fructose which prevents sucrose re-crystallization.[8] The –OH and–COOH groups of decomposed products promote binding of Mg^{2+} ions in homogeneous solution.

Subsequent heating at 80˚C for 2 hr on a hot plate stirrer let the Mg^{2+} ions react with sucrose completely. The solution was then heated in an electric oven at 200˚C for complete dehydration and changing into a viscous dark brownish gel. The polymeric network (gel) trapped colloidal silica. During additional heating, the obtained gel in the oven converted to black foamed mass. Finally the obtained mass was ground into powders and calcined in an electric furnace at 500, 600, 700, 800, 900, and 1000˚C for 3hr. The calcinations process decomposed polymeric matrix into gases such as CO_2 and H_2O and resulting in a large amount of released heat. These produced gases prevent from agglomeration of calcined powders.

During calcination, existed carbon in the black precursor powder was burned out and a white colour powder was formed. Fig. 1 shows detail of forsterite powder synthesis in a flow chart.

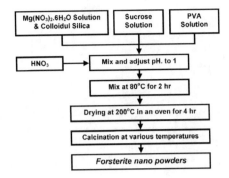

FIG. 1 Schematic flow chart of the synthesis of forsterite nanopowder.

Differential thermal analysis coupled with thermogravimetric (STA) was carried out on a Shimadzu STA-449C thermal analysis system in air atmosphere. X-ray diffraction (XRD) patterns were recorded from 5 to 80° (2θ) by a Joel 8030 diffractometer (Cu-K$_\alpha$ radiation). The apparent crystallite size of the forsterite powder was determined using the Scherrer equation:

$$\beta\ (2\theta) = k\lambda\,/\,[L.\cos\theta_0\ ]$$

Where λ is the wavelength (= 0.15406 nm), θ_0 is the Bragg angle, k is a constant (= 0.94), L is the apparent crystallite size, and the $\beta\ (2\theta)$ (in radians) is half-width of the diffraction line.[9]

Three diffraction peaks (211), (120), and (400), which have the advantage of being well separated and have high intensities, were chosen for the measurement.

FTIR spectra were carried out by a Shimadzu-FTIR spectrophotometer. More over a field emission electron microscope (FEM) (Zeiss-LEO) was used for microstructural observation.

RESULTS AND DISCUSSION

The STA curves of the precursor are shown in Fig. 2. The small endothermic peak about 110°C in DTA accounted for 6 % of the initial weight loss in TG, is assigned to exit of free water. As it can be seen 50% weight loss between 200 and 420 °C is due to decomposition of precursor and burning of carbon. Simultaneously the exothermic peak between 200 and 420°C proves the burning of polymeric matrix. Also the second exothermic peak (520°C) may be caused by the formation of periclase (MgO). The third broad exothermic peak (700-800°C) is due to forsterite formation.

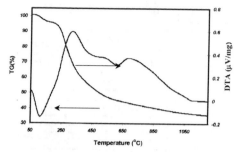

FIG. 2 STA curves of the dry forsterite precursor (heating rate=5°C/min).

The effect of calcining temperature on the formation of forsterite has been investigated by XRD. Fig. 3 illustrates X-ray diffraction patterns of as-received and calcined precursor at different temperatures. The precursor is essentially amorphous. Periclase has begun to form around 600°C, while forsterite formed between 700 to 800°C according to heterogeneous reaction between periclase and amorphous silica. With increasing temperature from 700 to 800°C a sudden increase in the intensity of forsterite reflections accompanied by sharpening of them occurred. Periclase was not detected in XRD pattern at higher temperatures than 800°C.

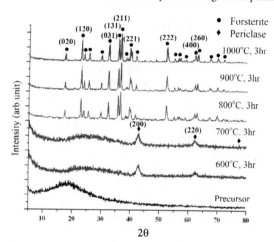

FIG. 3 X-ray diffraction patterns of as-rescieved and calcined precursor at different temperatures.

Table I shows the approximate crystallite size of forsterite calcined between 800 to 1000°C according to determination by the XRD line-broadening technique. The average crystallites size of calcined powder at 800°C is about 20 nm. Also according to FEM micrograph, the calcined powder at 800°C is agglomerated and its particle size is smaller than 150-200 nm.

TABLE I The approximate crystallite size and characteristics of the forsterite powder calcined at different temperatures

T(°C)	d$_{211}$(°A)	D (nm)	d$_{120}$(°A)	D (nm)	d$_{400}$(°A)	D (nm)
800/3hr	2.456	29	3.881	12	1.495	18
900/3hr	2.456	45	3.881	31	1.495	36
1000/3hr	2.456	61	3.881	48	1.495	52

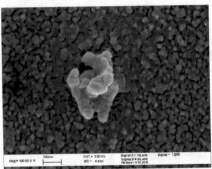

FIG . 4 FEM micrograph of the calcined forsterite at 800°C for 3hr.

The formation of forsterite from precursor has been followed using FT-IR. Fig. 5 illustrates FT-IR spectra of precursor and calcined powders at 730 and 800°C. FTIR analysis was used to follow the dehydration reaction of the precursor and the formation of Mg-O-Si bands. Common bands exist in precursor spectrum (Fig. 5,a) such as the broad OH band centred around 3400 cm^{-1}, the 1630 cm^{-1} H$_2$O vibration band, the bands related to NO$_3$ groups at 1384 cm^{-1}, the 1200-1400 cm^{-1} regions corresponding to C-H and C-O bands and also the peaks between 1000-1100 cm^{-1} is related to Si-O-Si bands. The decrease in intensities of mentioned bands and disappearance of them at 730°C (Fig. 5, b) is accompanied with the shift of Si-O-Si related peaks to lower wave number because of strengthening of Si-O bands in the SiO$_4$ tetrahedron that prove the formation of forsterite as it is shown in XRD of this sample. In calcined samples at 800°C (Fig. 5, c) the bands related to the characteristic peaks of forsterite appear in the range of 830-1000 cm^{-1} (SiO$_4$ stretching), at 500-620 cm^{-1} (SiO$_4$ bending) and at 475 cm^{-1} for modes of octahedral MgO$_6$. Similar results were also reported in previous studies.[10]

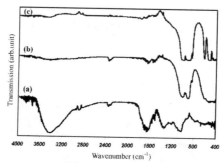

FIG. 5 FTIR spectrum of precursor (a), calcined precursor at 760°C (b) and calcined precursor at 800°C (c)

CONCLUSIONS

Nanocrystalline Forsterite (Mg$_2$SiO$_4$) has been synthesized by a polymer matrix method at low temperatures. The crystallite size is in the range of 10-30 nm, and its particle size is smaller than 200nm in the calcined sample at 800°C. This chemical synthesis through polymer matrix comprising sucrose and PVA, is applicable and cost effective compared to those of other processes, for the preparation of nano sized forsterite. According to XRD results a fully crystallized forsterite can be formed at 800°C and continues heating only has significant effect on the crystallites size and particle size of product.

REFERENCES

[1] J. H. Chesters, "Refractories production and properties", The Iron and Steel Institute, London (1973).

[2] M. I. Diesperova, V. A. Bron, V. A. Perepelitsyn, T. I. Boriskova, V. A. Alekseeva, E. I. Kelareva, "Forsterite Refractories from the Dunites of the Kytlym Deposits" Refractories and Industrial Ceramics, 18, No. 5-6, 278-282 (1977).

[3] N. I. Maliavski, O. V. Dushkin, J. V. Markina, G. Scarinci, "Forsterite Powder Prepared from Water-Soluble Hybrid Precursor", J. AIChE , 43, No. 11, 2832-36 (1997).

[4] C. Kosanovic, N. Stubicar, N. Tomasic, V. Bermanec, M. Stubicar, "Synthesis of a Forsterite Powder by Combined Ball Milling and Thermal Treatment", J. Alloys and Compounds, 389, 306-9 (2005).

[5] M.T. Tsai, "Effects of Hydrolysis Processing on the Character of Forsterite Gel Fibers. Part II: Crystallites and Microstructural Evolutions", J. Euro. Ceram. Soc., 22, No. 7, 1085-94 (2002).

[6] R. K. Pati, J. C. Ray, P. Pramanik, "A Novel Chemical Route for the Synthesis of Nanocrystalline α-Al$_2$O$_3$ powder", Materials letters, 44, No.5, 299-303 (2000).

[7] S.K. Mishra, S.K. Das P. Ramachandrarao, D.Yu Belov, "Synthesis of Zirconium Diboride-Alumina Composite by the Self-Propagating, High-temperature Synthesis Process", Metallurgical and Materials Transactions A, 34, No. 9.1979-82 (2003).

[8]J.C. Ray, A.B. Panda, P. Pramanik, "Chemical Synthesis of Nanocrystals of Tantalum Ion-doped Tetragonal Zirconia", Materials Letters, **53**, No.3, 145-50 (2002).
[9]B. D. Cullity, "Elements of X-ray Diffraction", Second edition, Addison-Wesley publishing Co., 281-5 (1978).
[10]M.T. Tsai, "Hydrolysis and Condensation of Forsterite Precursor Alkoxides: Modification of the Molecular Gel Structure by Acetic Acid", J. Non-crystalline Solids, **298**, No. 2-3,116-130 (2002).

DEFORMATION MECHANISMS OF NATURAL NANO-LAMINAR COMPOSITES: DIRECT TEM OBSERVATION OF ORGANIC MATRIX IN NACRE

Taro Sumitomo, Hideki Kakisawa
National Institute for Materials Science
1-2-1 Sengen, Tsukuba, Ibaraki 305-0047 Japan

Yusuke Owaki, Yutaka Kagawa
University of Tokyo
4-6-1 Komaba, Meguro, Tokyo 153-8904 Japan

ABSTRACT
Nacre is a natural nano-laminar composite making up the inner shell structure of mollusks. It consists of ordered plates of the $CaCO_3$ polymorph aragonite in an organic matrix, and known to have mechanical properties far greater than the monolithic ceramic. While making up less than 5% of the volume, the organic matrix in nacre is considered as playing a key role in the formation and mechanical behavior.

While the micro-scale mechanical behavior of the nacreous plate layers is well known, deforming by mechanisms common to composite materials such as sliding, pullout, crack deflection, and interface separation, the mechanical behavior of the matrix is still largely unclear. In order to further understand its nano-scale structure and deformation, transmission electron microscopy (TEM) was used to directly observe the structure and deformation behavior of the organic matrix.

Samples of nacre were partially demineralized to observe the structure of the organic material. The matrix between layers of aragonite plates (inter-layer matrix) was found to consist of perforated sheets with many holes, while the matrix between adjacent plates in a layer (inter-plate matrix) was relatively solid which suggested a material transport role of the inter-layer materix. Samples were also deformed in situ using a TEM straining holder to study the nano-scale deformation behavior of the organic matrix. Adhesion, bridging and high ductility of the organic material were observed, demonstrating its important role in toughening mechanisms.

INTRODUCTION

FORMATION OF NACRE
The characteristic structure of nacre has been described as "brick and mortar" [1] made up of ordered layers of aragonite plates separated by layers of organic matrix, illustrated in Fig. 1a-c. Nucleation and growth of nacreous layers are well discussed in the literature, where aragonite crystals have commonly been thought to grow epitaxially upon layers of organic material [2-4]. More recently, it has been suggested that the nucleation mechanism involves mineral bridges through pores in the organic matrix from one plate layer to the next [5].

A noted characteristic of nacre is that the plates are well aligned crystallographically in the [001], or c-axis direction. In the epitaxial nucleation model, this highly ordered structure is due to Ca^{2+} ions in the aragonite (001) plane preferentially bonding to proteins in the organic layer [6]; while nucleation via mineral bridges from one crystal to the next explains the uniform orientation in the mineral bridges model [5,7]. After nucleation, plates grow preferentially in the c-axis direction, hence plates grow first to their final thickness then laterally until a complete layer is formed [8-10]. It is generally agreed that the nucleation and growth process is controlled by proteins in the organic matrix which has two significant roles: to control the crystal orientation, morphology, and size during formation of aragonite from amorphous $CaCO_3$ [11]; and also to provide the physical structure for crystallization to occur.

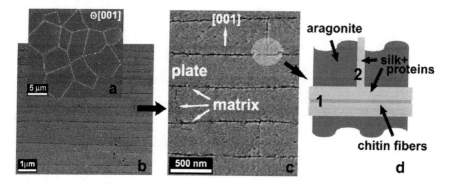

Figure 1. a. illustration of nacre surface structure, showing close packed plates within organic matrix. Examples of polygonal single plates are delineated by white lines and are approx. 5-8 μm wide. Plate surfaces uniformly perpendicular to the [001] direction (out of page). b and c. "brick and mortar" cross-section structure of nacre, consisting of horizontal layers of aragonite plates within an organic matrix. Single plates are uniformly aligned in the [001] direction, and approx. 500 nm thick. d schematic of the plate-matrix interface, showing: 1. inter-layer matrix, the organic material between plate layers, approx. 30 nm thick; 2. inter-plate matrix, the organic material between adjacent plates within a layer, approx. 10-20 nm thick

ORGANIC MATRIX STRUCTURE

Two types of organic matrix exist, the matrix between the layers of nacre plates, designated in this work as the inter-layer matrix, and the matrix between adjacent plates within a layer, the inter-plate matrix. This is illustrated in Fig. 1d. The inter-layer organic matrix makes up the bulk of the organic material, and has been described as sheets of aligned chitin fibers between silk and various proteins [1, 12]. The chitin at the core provides structural stability and a platform for crystallization [1, 13, 14]. The silk proteins have been observed as an unordered hydrated gel phase and reported to fill the spaces between the chitin [11, 15]. Proteins for nucleation, control of polymorphs, crystal orientation and restriction of growth have been reported [6, 8, 10, 13, 16-21]. The inter-plate matrix consists of organic material, largely silk with other proteins which were remained between the plates after completion of growth [9, 11].

Direct electron microscope observation of the organic material has been carried out by decalcification of nacre, revealing perforated sheets [2, 5, 20], and a similar structure has been observed after cleavage at the plate interface [22].

BULK MECHANICAL PROPERTIES

Bulk mechanical properties have been well investigated in the literature. Reported tests include tension/compression, shear, and indentation, from which common composite material mechanisms have been observed, such as sliding, pull-out, crack blunting/deflection, and crack bridging [22-27]. Strengthening and toughening mechanisms have also been described, which include nanoasperities on the surface which resist sliding [22, 28, 29], mineral bridges providing crack resistance [26]; crack deflection and deformation in the organic layer [23, 25, 27], adhesion and bridging of cracks by organic ligaments [22, 24, 27, 30] and plate interlocking [31, 32]. While the observed mechanisms show that the organic material plays a significant role in the bulk mechanical behavior, comparatively little is known about the mechanical behavior of the matrix itself.

ORGANIC MATRIX MECHANICAL PROPERTIES

Organic material adhesion to aragonite plates, fibers bridging cracks, and high ductility of the fibers observed after mechanical testing have been previously observed and reported as contributing to toughening mechanisms [27].

In this work, TEM was used to further study the structure and mechanical behavior of the organic matrix of nacre. In particular, nacre samples were partially demineralized to better observe the organic material structure, and a straining holder was used to observe the deformation behavior in situ.

EXPERIMENTAL METHOD

Nacre from abalone species haliotis gigantea was cut to size using a diamond circular blade. Grid mounted samples for normal TEM observation, and foil mounted samples for in situ straining were prepared. In the first case, cut sections were polished, mounted onto 3 mm diameter Mo grids, dimpled, and partially demineralized in 0.1M ethylenediaminetetraacetic acid (EDTA) for 30-60 min. Some of these were dehydrated in graded acetone, and fixed in 2% glutaraldehyde for 2 h. For the foil mounted samples, a notch approximately 200 μm wide was cut into a rectangular sample about 4 mm long and 2 mm wide, which was then polished and mounted onto a straining holder Cu foil. This was ground and polished further, dimpled and partially demineralized with 0.1M EDTA for 2 h.

The grid mounted samples were observed in a JEOL 2010 TEM at 200 kV, the foil mounted samples were observed in a JEOL 2000FX TEM with a JEOL EM-SEH3 straining holder and CCD movie recording system at 200 kV. Straining rate was approximately 50-100 nm/s.

RESULTS AND DISCUSSION

ORGANIC MATRIX STRUCTURE

Figure 2 shows overlapping sheets of inter-layer organic matrix after demineralization in 0.1 M EDTA for 1 hour and fixing in 2% glutaraldehyde. The most notable features are the holes, which are about 10-100 nm in diameter and randomly distributed. This perforated structure matches previous electron microscopy observations [2, 20]. Figure 3 shows inter-plate organic matrix stretched between separated plates from a foil mounted sample. Fig. 3a shows a still image taken from movie footage during deformation. 3b shows a film micrograph after further deformation. The inter-plate organic material was typically observed as solid sheets or fibers which stretched with plate separation.

Both types of organic matrix demonstrate structural stability. In the case of the inter-layer matrix, this can be attributed largely to the ordered chitin fibers. In the case of the inter-plate matrix, this stability demonstrates that silk proteins also provide some order [14]. The most notable difference between the two is the large numbers of holes in the inter-layer organic sheets. This supports reports in the literature which suggest the holes provide transport paths for nucleation and growth of crystals. either directly in the form of mineral bridges [5]; or as locations for membranes which allow passage of amorphous $CaCO_3$ [33].

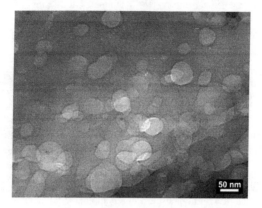

Figure 2. TEM micrograph of overlapping organic matrix sheets from the inter-layer interface.

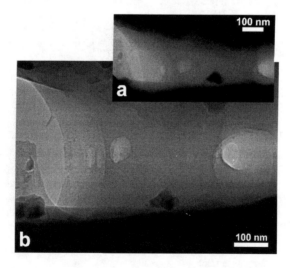

Figure 3. TEM micrographs showing inter-plate organic matrix stretched between adjacent plates at various stages in separation. a Movie still image after some extension. b. Scanned film micrograph after further extension.

ORGANIC MATRIX DEFORMATION

Figures 4 and 5 present still images of inter-plate matrix deformation where a TEM straining holder was used for in situ deformation. It is noted that the sample initially failed at low strain in a brittle manner, and further deformation could not be directly controlled. However post-fracture separation of plates allowed the direct movie recording of deformation and failure of the organic matrix.

Figure 4 shows fibers adhered to the sides of the plates which extend as the plates separate. Necking type failure is observed as matrix fibers stretch, become thinner, and break. Figure 5 shows similar type of failure in thinner fibers. Assuming an approximate initial length of 20 nm measured from the inter-plate separation of undeformed plates, elongation of organic fibers at failure can be measured, and range from about ten to fifteen times the initial length. This compares reasonably well with the twenty times elongation reported in other work [30]. The adhesion and ductility demonstrate the significant contribution to bulk toughening by the organic material.

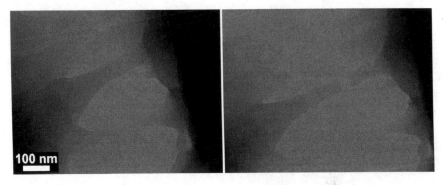

Figure 4. TEM still images showing deformation of inter-plate organic matrix fibers as adjacent plates separate. Fibers are seen to stretch and break.

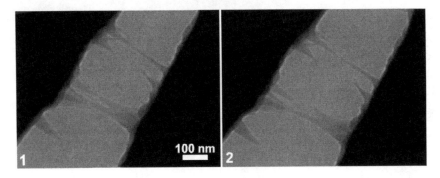

Figure 5. TEM still images showing deformation of thin organic fibers.

CONCLUSIONS

TEM observation of demineralized nacre of abalone *haliotis gigantea* showed a perforated sheet structure of the inter-layer matrix and sheets and fibers of inter-plate matrix. The presence of many holes in the former tend to support theories of crystal nucleation and growth where the holes were used as transport pathways between layers of organic matrix.

A straining holder was used to make in situ TEM observations of the deformation behavior of the inter-plate organic material. Adhesion and bridging of organic sheets and fibers with high ductility demonstrated significant contribution to bulk toughening.

While the organic matrix in nacre makes up less than 5% by volume of the whole, it has been shown that it plays an important and complex role in not only in the formation of nacre structure, but also in its mechanical properties.

REFERENCES

[1]M. Sarikaya, "An introduction to biomimetics: a structural viewpoint", *Microsc. Res. Tech.*, **27**, 360-375 (1994).

[2]K. M. Towe and G. H. Hamilton, "Ultrastructure and inferred calcification of the mature and developing nacre in bivalve mollusks", *Calc. Tiss. Res.*, **1**, 306-318 (1968).

[3]G. Bevelander and H. Nakahara, "An electron microscope study of the formation of the nacreous layer in the shell of certain bivalve molluscs", *Calc. Tiss. Res.*, **3**, 84-92 (1969).

[4]G. Bevelander and H. Nakahara, "An electron microscope study of the formation and structure of the periostracum of a gastropod, Littorina littorea", *Calc. Tiss. Res.*, **5**, 1-12 (1970).

[5]T. E. Schäffer, C. Ionescu-Zanetti, R. Proksch, M. Fritz, D. A. Walters, N. Almqvist, C. M. Zaremba, A. M. Belcher, B. L. Smith, G. D. Stucky, D. E. Morse and P. K. Hansma, "Does abalone nacre form by heteroepitaxial nucleation or by growth through mineral bridges?" *Chem. Mater.*, **9**, 1731-1740 (1997).

[6]S. Weiner and W. Traub, "Macromolecules in mollusc shells and their functions in biomineralization", *Phil. Trans. R. Soc. Lond. B*, **304**, 425-434 (1984).

[7]A. G. Checa and A. B. Rodríguez-Navarro, "Self-organisation of nacre in the shells of Pterioida (Bivalvia: Mollusca)", *Biomaterials*, **26**, 1071-1079 (2005).

[8]A. Lin and M. A. Meyers, "Growth and structure in abalone shell", *Mater. Sci. Eng., A*, **390**, 27-41 (2005).

[9]H. Nakahara, "An electron microscope study of the growing surface of nacre in two gastropod species, Turbo cornutus and Tegula pfeifferi", *Venus (Jap. Jour. Malac.)*, **38**, 205-211 (1979).

[10]M. Rousseau, E. Lopez, A. Couté, G. Mascarel, D. C. Smith, R. Naslain and X. Bourrat, "Sheet nacre growth mechanism: a Voronoi model", *J. Struct. Biol.*, **149**, 149-157 (2005).

[11]L. Addadi, D. Joester, F. Nudelman and S. Weiner, "Mollusk shell formation: a source of new concepts for understanding biomineralization processes", *Chem. Eur. J.*, **12**, 980-987 (2006).

[12]S. Weiner and L. Addadi, "Design strategies in mineralized biological materials", *J. Mater. Chem.*, **7**, 689-702 (1997).

[13]T. Kato, A. Sugawara and N. Hosoda, "Calcium Carbonate-Organic Hybrid Materials", *Adv. Mater.*, **14**, 869-877 (2002).

[14]S. Weiner and W. Traub, "X-ray diffraction study of the insoluble organic matrix of mollusk shells", *FEBS Lett.*, **111**, 311-316 (1980).

[15]Y. Levi-Kalisman, G. Falini, L. Addadi and S. Weiner, "Structure of the nacreous organic matrix of a bivalve mollusk shell examine in the hydrated state using Cryo-TEM", *J. Struct. Biol.*, **135**, 8-17 (2001).

[16]A. M. Belcher, X. H. Wu, R. J. Christensen, P. K. Hansma, G. D. Stucky and D. E. Morse, "Control of crystal phase switching and orientation by soluble mollusc-shell proteins", *Nature*, **381**, 56-58 (1996).

[17]J. L. Arias and M. S. Fernández, "Biomimetic processes through the study of mineralized shells", *Mater. Charact.*, **50**, 189-195 (2003).

[18]C.-S. Choi and Y.-W. Kim, "A study of correlation between organic matrices and nanocomposite materials in oyster shell formation", *Biomaterials*, **21**, 213-222 (2000).

[19]Q. L. Feng, G. Pu, Y. Pei, F. Z. Cui, H. D. Li and T. N. Kim, "Polymorph and morphology of calcium carbonate crystals induced by proteins extracted from mollusk shell", *J. Cryst. Growth*, **216**, 459-465 (2000).

[20]C. Grégoire, "Structure of the Molluscan Shell", in M. Florkin and B. T. Scheer, eds., *Mollusca*, New York, 45-102 (1972).

[21]X. Shen, A. M. Belcher, P. K. Hansma, G. D. Stucky and D. E. Morse, "Molecular cloning and characterization of lustin A, a matrix protein from shell and pearl nacre of Haliotis rufescens", *J. Biol. Chem*, **272**, 32472-32481 (1997).

[22]R. Z. Wang, Z. Suo, A. G. Evans, N. Yao and I. A. Aksay, "Deformation mechanisms in nacre", *J. Mater. Res.*, **16**, 2485-2493 (2001).

[23]J. D. Currey, "Mechanical properties of mother of pearl in compression", *Proc. R. Soc. Lond. B.*, **196**, 443-463 (1977).

[24]A. P. Jackson, J. F. V. Vincent and R. M. Tuner, "The mechanical design of nacre", *Proc. R. Soc. Lond. B.*, **234**, 415-440 (1988).

[25]R. Menig, M. H. Meyers, M. A. Meyers and K. S. Vecchio, "Quasi-static and dynamic mechanical response of Haliotis rufescens (abalone) shells", *Acta. Mater.*, **48**, 2383-2398 (2000).

[26]F. Song, A. K. Soh and Y. L. Bai, "Structural and mechanical properties of the organic matrix layers of nacre", *Biomaterials*, **24**, 3623-3631 (2003).

[27]R. Z. Wang, H. B. Wen, F. Z. Cui, H. B. Zhang and H. D. Li, "Observations of damage morphologies in nacre during deformation and fracture", *J. Mater. Sci.*, **30**, 2299-2304 (1995).

[28]F. Barthelat, C.-M. Li, C. Comi and H. D. Espinosa, "Mechanical properties of nacre constituents and their impact on mechanical performance", *J. Mater. Res.*, **21**, 1977-1986 (2006).

[29]A. G. Evans, Z. Suo, R. Z. Wang, I. A. Aksay, M. Y. He and J. W. Hutchinson, "Model for the robust mechanical behaviour of nacre", *J. Mater. Res.*, **16**, 2475-2484 (2001).

[30]B. L. Smith, T. E. Schäffer, M. Viani, J. B. Thompson, N. A. Frederick, J. Kindt, A. M. Belcher, G. D. Stucky, D. E. Morse and P. K. Hansma, "Molecular mechanistic origin of natural adhesives, fibres and composites", *Nature*, **399**, 761-763 (1999).

[31]F. Barthelat, H. Tang, P. D. Zavattieri, C.-M. Li and H. D. Espinosa, "On the mechanics of mother-of-pearl: a key feature in the material hierarchical structure", *Journal of the Mechanics and Physics of Solids*, **In press**, (2006).

[32]K. S. Katti, D. R. Katti, S. M. Pradhan and A. Bhosle, "Platelet interlocks are the key to toughness and strength in nacre", *J. Mater. Res.*, **20**, 1097-1100 (2005).

[33]H. Mutvei, "Ultrastructure of the mineral and organic components of molluscan nacreous layers", *Biomineralization*, **2**, 48-61 (1970).

PHOTOCATALYTICALLY SENSITIVE MATERIALS FOR WATER SPLITTING

Marek Laniecki

Faculty of Chemistry, A. Mickiewicz University
Grunwaldzka 6
60-780 Poznan, Poland

ABSTRACT

An overview of the semiconducting materials applied in photocatalytic water splitting, operating both under UV and visible radiation, is given. The major part concentrate on different methods of pure titania preparation and their modification with sulfur, nitrogen or carbon. The modification of titania with noble metals, anions and dye sensitization is also described. Other semiconducting photocatalysts such as sulfides, perovskites, nitrided compounds of tantalum, indium or vanadium and their application in photocatalytic splitting of water are also discussed.

INTRODUCTION

The world is at the treshold of a new economy which will require not only the use of new energy resources but also new energy carriers which will be environmentally benign, energy efficient, "multipurpose" and produced in large quantities at low prices. One intended alternative to fossil fuels is hydrogen which fulfill all these requirements [1,2].

Hydrogen is widely acknowledged as as an energy-efficient and eco-friendly fuel of the future. Conventional production processes of hydrogen are based on fossil fuels and are not only energy intensive but also incur high production costs. Therefore, production of hydrogen using alternative routes has drawn the attention in many research groups. Among various future methods of hydrogen generation the photocatalytic splitting of water [3] as well as application of microorganisms in this process („dark" or „illuminated" fermentation) seem to be very important methods [4].

Photocatalytic splitting of water is one of the proposed methods which can be applied in hydrogen generation from water. In this process the n-type semiconductor under illumination generates pairs of electrons and positively charged holes with energy greater or equal to the bandgap (E_{bg}). Photogenerated electrons migrate to the surface of semiconductor and cause photoreduction of water, whereas positively charged holes oxidize water with subsequent oxygen evolution.

In 1972 Fujishima and Honda [5] discovered the photocatalytic splitting of water applying rutile (TiO_2) as photoanode and platinum as catode. Since that time numerous publications related with basic research in the field of photocatalytic hydrogen generation, photooxidation of organic pollutants and application of photocatalysts in removal of inorganic wastes appeared in literature [3,6].

Titanium oxide is the most studied semiconductor in photocatalysis and majority of light driven reactions, eg. photodegradation of organic pollutants in mineralization process, was performed with this more or less modified oxide. Application of TiO_2 as the model photocatalyst allowed to establish certain rules governing the photocatalytic processes.

The main disadvantage of application of titania as photocatalyst in large scale is related with its relatively high value of band gap (3.2 eV). This value shows that application of pure TiO_2

(anatase) in photogeneration of hydrogen is limited to the UV radiation and can be effective only below 382 nm. Since UV radiation of solar spectrum do not exceed 4%, the use of pure titania in photocatalytic splitting of water should be limited only to the basic research. From the viewpoint of efficient solar energy utilization it is necessary to develop efficient photocatalysts that work under visible light (50% of solar spectrum). Until recent years there were no known effective photocatalysts for water splitting operating under visible irradiation. Recent studies shown, however, that photocatalytic water splitting under visible light is possible. Although the activities of these systems are still far beyond the expectations and inadequate for practical application in large scale it is belived that further studies will solve the problem "solar" water splitting.

BASIC PRINCIPLES OF SEMICONDUCTOR CATALYSIS

Considering electronic structure, each semiconductor of the chalcogenide type consists of valence band (VB) and conduction band (CB). Energy difference between these two bands is defined as an energy band gap (E_{bg}). When a semiconducting photocatalysts is illuminated with photons whose energy is equal or higher than their band gap, there is an adsorption of these photons with simultaneous creation of electron-hole pairs within the bulk of solid. The electrons migrate in very rapid excitation process (usually less than few picoseconds) towards conduction band (CB) and remain there in excited state for few microseconds, whereas postively charged holes remain in the valence band (VB). Time of excitation of few microseconds is long enough to perform the photocatalytic reaction. Here, in the presence of gaseous or liquid reagents the spontaneous adsorption occurs and according to the redox potential of each adsorbate an electron transfer proceeds towards acceptor molecule, whereas positively charged holes are transferred to donor molecules.

The potential level of acceptor species is thermodynamically required to be below the potential of CB of semiconductor, whereas donor level must be located above the VB and therefore donate electron to a positively charged hole. In all aqueous solutions the pH value strongly influences the band edge positions of different semiconductors.

The efficiency of the photocatalytic process is measured as quantum yield (number of events per photon adsorbed) and in many cases is significantly reduced by recombination of photogenerated CB electrons and positively charged VB holes. Therefore, charge separation is one of the most important tasks that should be solved by the choice and appropriate enengineering of photocatalysts.

PHOTOCATALYTIC SPLITTING OF WATER OVER TiO$_2$

At pH= 0 the redox potential of H$^+$/H$_2$ and O$_2$/OH$^-$ equals 0 and 1.23 eV, respectively. The overall splitting of water over semiconducting photocatalysts can be performed under certain condition: a bottom of CB must be more negative than H$^+$/H$_2$ redox potential and O$_2$/OH$^-$ at VB has to be more positive. This condition is easily fulfilled by TiO$_2$ (anatase)- the most widely used photocatalyst. The use of this photocatalyst in large scale has, however, several serious drawbacks. This concern also hydrogen generation *via* photocatalytic splitting of water. Due to the high value of energy band gap of titania (E_{bg} = 3.2 eV) photocatlytic systems can operate only within the UV region of solar radiation (λ < 382 nm). Although the lifetime of photogenerated electron/holes pairs is relatively long, the efficiency of photogenerated hydrogen is very low, due to the reverse reaction: recombination between hydrogen and oxygen and water formation.

In order to overcome these obstacles several actions can be undertaken. Since the early eighties several groups proposed different solutions. The selection of appropriate methods of titania synthesis is the most important one [7,8]. The modification of chemical properties of TiO_2 based materials by affecting both structural and electronic properties is crucial to obtain a new photocatalytically active anatase structures.

There are three ways of preparation of novel TiO_2 based materials: formation of nanotubes, mesoporous grain powders and TiO_2-containing materials of the MCM-41 or SBA-15type.   Kasuga [9] and other papers [10,11] reported recently the new approaches that allow to control two main properties: the inner shell diameter of the nanotube and chemical composition (presence of H in addition to Ti and O) of the nanotube walls. Nanotubes are usually obtained by basic attack (NaOH) to TiO_2 precursor solid materials an subsequent rinse with acids and distilled water. Control of the inner shell diameter is accomplished by varying the TiO_2 to NaOH molar ratio [10], while the adequate anatase-type structure can obtained by controlling the pH at the rinsing steps and managing the structural and morphological properties of titania presursor[11]. Additionally, the use of AAO (anodic aluminium oxide) templates could give enough flexibility to obtain hierarchically ordered arrays of nanotubes with varing wall thickness and also to synthesise oxide-oxide nanocomposites [12].

Template directed techniques using surfactants (different types of Pluronic, polyethylene oxide) are used to obtain hierarchically ordered mesoporous titania by sol-gel [16] and microemulsion procedures [15,18], but it is still an open question how to obtain material with high crystallinity and low defect density[13-17]. The collapse of the porous structure during thermal treatments apperas to be limited by final calcination temperature and heating rate [14], whereas the thickness of the walls can controll crystallinity [15,16] of TiO_2. Similar effects are observed while applying cations as the doping agents of TiO_2 [17].

The MCM-41 or SBA-15 mesoporous materials can be modified by titanium ions either by incorporation into the walls [18] or by grafting or deposition of TiO_2 inside the channels [19-22].

The improvement of the yield of photocatalytically generated hydrogen can be also achieved by addition of electrons donors (oxygen scavengers), noble metal deposition, metal ion doping, anion doping or sensitization of TiO_2 with dyes.

Different organic compounds has been already applied as the electron donors while using titania as photocatalyst. Mainly methanol, ethanol , EDTA or formaldehyde were applied as the effective electron donors [23-25]. These reagents, also known as sacrificial agents or oxygen scavengers, are oxidized by VB holes whereas the stream of newly generated electrons reduces protons to molecular hydrogen. Moreover, due to the addition of sacrificial agents it is reduced significantly chance of reverse reaction between H_2 and O_2.

For certain semiconductors couples of inorganic ions can be applied as sacrificial agent, as well. For example S^{2-}/SO_3^{2-} system was applied when CdS photocatalysed water splitting [26]. The redox IO_3^-/I^- couple was tested with the system that mimics natural photosynthesis [27-30]. The main idea of this system operating under visible light, also called „Z-scheme", is based on application of two photocatalysts. System composed from WO_3 and doped $SrTiO_3$ containing redox couple mediator can generate hydrogen and oxygen without consumption of sacrificial reagent. Here, for WO_3 photocatalyst, photoelectrons reduce IO_3^- ions to I^- and positive holes oxidize water into O_2, whereas photogenerated holes in doped $SrTiO_3$ oxidize I^- to IO_3^- with simultaneous reduction of H^+ to molecular hydrogen. The Z- scheme is an example of two step photoexcitation process that can work under natural irradiation.

Although pure titania is effective photocatalyst only in UV light, recent years proved that this shortcoming can be overcome by doping of pure titania with such elements like nitrogen [31-37], carbon [38,39], sulfur [40-42], phosphorous , fluorine etc. The doping effect with non-metal elements, in contrast to cation doping, is based on formation of lower amounts of recombination centers and therefore can enhance photocatalytic activity. Moreover, the newly formed impurity states are near the VB edge, but do not act as charge carriers. Doping of titania with C or N results in approximately 50 nm red shift in absorption spectra and in consequence better photoactivity in the visible. Recent results with carbon doping of titania proved better activity in water splitting [43]. In all cases (C, N, S) mixing of p states (eg. p for N, 3p for S) with 2p states of oxygen shifts the VB upwards that results in narrowing of band gap for TiO_2.

Since many years the extensive research on metal ion doping effects on the photocatalytic activity (also in water splitting reaction) has been carried out [44-46]. Both transition metal ion and rare earth metal ions were studied. It was established that metal ions incorporated into the lattice of TiO_2 can expand the photoresponse into the visible, due to the shifts in VB and CD energy levels. However, the negative effect related with fast electron/hole recombination can diminish the photoefficiency of metal ion doped titania. It was found that among 21 [47] studied ions only doping with Fe, Cu, Mn, Mo, V, Ru, Rh and Os ions resulted in the improved photoactivity. In contrast, there are a group of cations that speeds up the electron/hole recombination, decreasing simultaneously photocatalytic activity. Doping with Co, Cr, Al, as well as application of very high concentrations of the all doping ions are detrimental [48]. Studies performed with lanthanide cations [24,49] as dopants, revealed that gadolinum ions were the most effective in photocatalytic performance. This effect, among other lanthanide cations, resulted from the best ability in transferring charge carriers to the surface of TiO_2 particles. Recent paper by Peng et al. [50] in which berrylium ion was applied as TiO_2 dopant confirm earlier reasults that an external dopings is beneficial in hydrogen generation, whereas deep location of Be^{2+} ions within TiO_2 particles, results in fast electron/hole recombination and poor performance.

DYE-SENSITIZATION

The relatively large band gap in TiO_2 (3.2 eV) does not allow for the excitation of electrons with light wavelength higher than 382 nm. Moreover, very rapid recombination of electrons with VB holes in non-modified titania makes the yield of photogenerated hydrogen extermely low. The solution lays in dye sensitization of large band gap semiconductor. Under the visible light, the excited dye inject electrons to the CB of semiconductor and perform the reduction of water originated protons to molecular hydrogen [51,52]. Data presented in the paper by Jana [53] provide wavelength maxima of some of frequently used dyes. The sensitization with dyes require, however, their regeneration. For these purposes the redox couple such as IO_3^- /I or sacrificial agent (eg. methanol or EDTA) must be added to the reaction solution to sustain the hydrogen evolution. The yield of the photogenerated hydrogen is limited by the rate of the electron/hole recombination [54,55]. The shorter time of dye excitation and slower recombination time, the higher is the yield of photocatalyzed hydrogen.

OXIDE AND PEROVSKITE TYPE PHOTOCATALYSTS

Although titanium dioxide, including the most popular photocatalyst P-25 from Degussa (mixture of approximately of 80 % of anatase and 20 % of rutile) is one of the most popular and commercially available photocatalyst, other oxides alone or as composite semiconductors [56,57] can be apllied in photocatalytic splitting of water. The application of such oxides as ZnO, SnO_2 , CeO_2, ZrO_2, Nb_2O_5, Ta_2O_5 [58-62] in photocatalytic splitting of water proved that depending of the band gap of particular oxide the hydrogen „photoevolution" is possible and that reaction conditions (deposition of noble metals, use of different sacrificial agents, dye-senitization, application of basic pH, etc.) influences the amount of H_2.

In recent years appeared a number of papers describing pervskites and more or less complex oxide systems for water splitting and operating under UV illumination with relatively high quantum yield, eg. $SrTiO_3$ $K_2La_2Ti_3O_{10}$, $NiO/NaTaO_3$, containing transition metal cations with d^0 electronic configuration (Ti^{4+} Zr^{4+}, Nb^{5+},Ta^{5+}, W^{6+}) [63,64]. Moreover, mixed oxide photocatalysts with d^{10} electronic configuration $(Ga^{3+}, In^{3+}, Ge^{4+}, Sn^{4+}$ and $Sb^{5+})$ has been recently identified as the efficient photocatalysts for the overall splitting reaction [65]. $BaIn_2O_4$, Ca_2SnO_4, $LiSbO_3$, Zn_2GeO_4 and $SrGa_2O_4$ are the examples of the family of stable and efficient photocatalysts for water splitting. Although compounds containing transition metal cations of d^0 and d^{10} configuration indicate good photoresponse they do not operate under visible irradiation. This result in continous search for appropriate modification of these compounds in order to narrow the band gaps and to slow down the recombination of photoexcited electron/hole pairs.

NON-OXIDE PHOTOCATALYSTS FOR OVERALL WATER SPLITTING

In an overview paper Ashokkumar [66] describes different semiconductor pariculate systems for photocatalytic hydrogen generation, and among the others, CdS is proposed as the very efficient one. Cadmium sulfide with the band gap of 2,4 eV appears to be an ideal candidate for this reaction, especially that it can operate under visible irradiation [67]. However, the very high initial yield of photogenerated hydrogen [68,69] decreases rapidly due to photocorrosion. Similar effects were observed with other oxides , eg . ZnS or Bi_2S_3 [70,71].

Until several years ago, it was belived that non-oxide semiconductors operating in the visible can not be successfuly applied in photocatalytic splitting of water due to their poor stability and rapid deactivation. Recent years, however, revealed that certain nitrides or oxynitrides of the transition metals with the d^0 and d^{10} electronic configuration can be efficient photocatalysts in water splitting [72,73]. Domen [74] describes Ta_3N_5 and TaON, $BaTaO_2N$, as well as systems containing GaN or different crystallographic phases of Ge_3N_4 as the photocatalysts with band gap lower than 3 eV operating under visible light. Although the efficiency of photocatalytic hydrogen with nitrides or oxynitrides is rather low, due to large amount of crystal defects, it is belived that improvement of the synthesis of these compounds will provide good catalysts for water splitting.

REFERENCES

1. T. N. Veziroglu, F. Barbir, Hydrogen Energy Technolgies, UNIDO Emerging Technology Series, UNIDO, Vienna , 1998.
2. C.-J. Winter, Int. J. Hydrogen Energy, 29 (2004) 1095.
3. A. L. Linsebigler, G. Lu, J. T. Yates Jr., Chem. Rev. 95 (1995) 735.

4. D. Das, T. N. Veziroglu, Int. J. Hydrogen Energy, 26 (2001) 13.
5. A. Fujishima, K. Honda, Nature, 238 (1972) 37.
6. A. Mills, S. Le Hunte, J. Photochem. Photobiol. A: Chem., 108 (1998)1.
7. M. Fernandez-Garcia, A. Martinez-Arias, J. C. Hanson, J. A. Rodriguez, Chem. Rev., 104 (2004) 4063.
8. A. S. Barnard, L. A. Curtiss, Nanoletters, 5 (2005)126.
9. T. Kasuga, Thin Solid Films, 496 (2006) 141.
10. D. V. Bavykin, V. N. Parmon, A. A. Lapkin, F. C. Walsh, J. Mater. Chem. 14 (2004) 3370.
11. X. Wu, Q.-Z. Jian, Z.-F. Ma, M. Fu, W.-F. Shangguan, Solid State Comm.,136 (2005) 513
12. L. Cheng, X. Zhang, B. Liu, H. Wang, Y. Li, Y. Huang, Z. Du, Nanotechnology, 16 (2005) 1341.
13. J. J. Shye, M. R. de Guire, J.Am.Chem.Soc., 127 (2005) 12736.
14. B. L. Krisch, E. K. Richman, A. E. Riley, S. H. Tolbart, J. Phys. Chem. B, 108 (2004) 12698.
15. M. Fernandez-Garcia, C. Belver, X. Wang, J. Hanson, J. A. Rodriguez, J. Am. Chem. Soc. , in press
16. Y. Sakatani, D. Grosso, L. Nicole, C. Biossiere, G. J. A. Soler-Illia, C. Sanchez, J. Mater. Chem., 16 (2006) 77.
17. J. N. Kondo, T. Yamashita, K. Nakajima, D. Lu, M. Hara, K. Domen, J. Mater. Chem. 15 (2005) 2035.
18. A. Corma, Chem. Reviews, 97 (1997) 2373.
19. E. P. Reddy, B. Sun, P. G. Smirniotis, J. Phys. Chem., 108 (2004)17198.
20. M. Alvaro, E. Carbonell, V. Fornes, H. Garcia, ChemPhysChem, 7 (2006) 200.
21. M. Laniecki, M. Wojtowski, Stud. Surf. Sci. Catal., 158 (2005) 757.
22. M. Ruszel, B. Grzybowska, M. Laniecki, M. Wojtowski, Catal. Comm. in press 2007.
23. N. L. Wu, M. S. Lee, Int. J. Hydrogen Energy, 29 (2004) 1601.
24. M. Zalas, M. Laniecki, Solar Energy Mat. & Solar Cells, 89 (2005) 287.
25. G. R. Bamwenda, S. Tsubota, T. Nakamura, M. Haruta, J. Photochem. Photobiol. A: Chem., 89 (1995) 177.
26. A. Koca, M. Sahin, Int. J. Hydrogen Energy, 27 (2002) 363.
27. K. Sayama, K. Mukasa, R. Abe, Y. Abe, H. Arakawa, Chem. Comm., (2001) 2416.
28. R. Abe, K. Sayama, K. Domen, H. Arakawa, Chem. Phys. Lett. 344 (2001) 339.
29. K. Sayama, K. Mukasa, R. Abe, Y. Abe, H. Arakawa, J. Photochem. Photobiol. A: Chem., 148 (2002) 71.
30. K. Lee, W. S. Nam, G. Y. Han, Int. J. Hydrogen Energy, 29 (2004) 1343.
31. R. Asahi, T. Morikawa, T. Ohwaki, K. Aoki, Y. Taga, Science, 293 (2001) 269.
32. H. Wang, J. P. Lewis , J. Phys.: Condens. Matter, 18 (2006) 421.
33. S. Sakthivel, H. Kisch, ChemPhysChem, 4 (2003)487.
34. S. Sakthivel, M. Janczarek, H. Kisch, J. Phys. Chem. B, 108 (2004) 19384.
35. H. Irie, Y. Watanabe, K. Hashimoto, J. Phys. Chem. B, 107 (2003) 5483.
36. R. Nakamura, T. Tanaka, Y. Nakato, J. Phys. Chem. B, 108 (2004) 10617.
37. T. Ihara, M. Miyoshi, Y. Iriyama, O. Matsumoto, S. Sugihara, Appl. Catal. B: Environmental 42 (2003) 403.
38. C. Lettman, K. Hildebrandt, H. Kisch, W. Macyk, W. F. Maier, Appl. Catal. B,

32 (2001) 215.

39. S. Sakthivel, H. Kisch, Angew. Chem. Int. Ed. Engl. 42 (2003) 4908.

40. T. Ohno, M. Akiyoshi, T. Umebayashi, K. Asai, T. Mitsui, M. Matsumura, Appl. Catal.A: General 265 (2004) 115.

41. T. Umebayashi, T. Yamaki, S. Tanaka, K. Asai, Chem. Lett. 32 (2003) 330.

42. T. Ohno, T. Mitsui, M. Matsumura, Chem. Lett. 32 (2003) 364.

43. S.U.M. Khan, M. Al-Sahry, W. B. Ingler, Science, 297 (2002) 2243.

44. M. Gratzel, R. F. Howe, J. Phys. Chem., 94 (1990) 2566.

45. M. Fujihara, Y. Satoh, T. Osa, Bull. Chem. Soc. Japan, 55 (1982) 666.

46. E. C. Butler, A. P. Davis, J. Photochem. Photobiol., A: Chem. 70 (1993) 273.

47. W. Y. Choi, M. R. Hoffmann, J. Phys. Chem., 84 (1994) 13669.

48. J.-M. Herrmann, J. Disdier, P. Pichat, Chem. Phys. Lett. 108 (1984) 618.

49. A. W. Xu, Y. Gao, H. Q. Liu, J. Catal. , 207(2002) 151.

50. S. Q. Peng, Y.X. Li, G. X. Lu, S. B. Li, Chem. Phys. Lett., 398 (2004) 235.

51. Z. C. Bi, H. T. Tien, Int. J. Hydrogen Energy, 9 (1984) 717.

52. R. Abe, K. Sayama, H. Arakawa, Chem. Phys. Lett. 362 (2002) 441.

53. A. K. Jana, J. Photochem. Photobiol. A: Chem, 132 (2000) 1.

54. S. G. Yan, J. T. Hupp, J. Phys. Chem. 100 (1996) 6867.

55. J. M. Rehm, G. L. Mclendon, Y. Nagasawa, K. Yoshihara, K. Moser, M. Gratzel, J. Phys. Chem. , 100 (1996) 9577.

56. K. Gurunathan, P. Maruthamuthu, V. C. Sastri, Int. J. Hydrogen Energy, 22 (1997) 57.

57. T.V. Nguyen, S. S. Kim, O. B. Yang, Catal. Comm. 5 (2004) 59.

58. K. Sayama, H. Arakawa, J. Photochem. Photobiol. A: Chem. 77 (1994) 243.

59. K. Sayama, H. Arakawa, J. Photochem. Photobiol. A: Chem. 94 (1996) 67.

60. M. S. Wrighton, Acc. Chem. Res., 12(1979) 303.

61. D. E. Aspnes, K. Heller, J. Phys. Chem. 87 (1983) 4919.

62. K. Nakajima, D. Lu, M. Hara, K. Domen, J. N. Kondo, Stud. Surf. Sci. Catal.,158 (2005) 1477.

63. H. Kato, K. Asakura, A.Kudo, J. Am. Chem. Soc. , 125 (2003)3082

64. M. Laniecki, R. Glowacki, Proceed. 15th WHEC, Yokohama 2004, CD-ROM edition, 7 pages.

65. J. Saito, N. Saito, H. Nishiyama, Y. Inoue, J. Photochem. Photobiol. A : Chem. 158 (2003) 139.

66. M. Ashokkumar, Int. J. Hydrogen Energy, 23 (1998) 427.

67. T. Oncescu, M. Contineanu, L. Meahcov, Int. J. Photoenergy, 1 (1999) 1.

68. N. Buchler, J. F. Reber, K. Meier, J. Phys. Chem., 88 (1984) 3261.

69. J. F. Reber, M. Rusek, 90 (1986) 824.

70. J. F. Reber, K. Meier, J. Phys. Chem. 88 (1984) 5903.

71. Y. Bessekhouad, M. Mohammedi, M. Trari, Sol. Ener. Mater. & Solar Cells, 7 (2002) 339.

72. M. Hara, G. Hitoki, T. Takata, J. N. Kondo, H. Kobayashi, K. Domen, Catal. Today, 78 (2002) 555.

73. A. Kasahara, K. Nukumizu, G. Hitoki, T. Takata, J. N. Kondo, M. Hara, K. Domen, J. Phys. Chem.A, 106 (2002) 6750.

74. K. Domen, Proceed. 15th WHEC, Yokohama 2004, CD-ROM edition, 12 pages.

BIOMIMETIC SYNTHESIS OF HIERACHICALLY POROUS MATERIALS AND THEIR STABILIZATION EFFECTS ON METAL NANOPARTICLES

Junhui He and Shuxia Liu
Functional Nanomaterial Laboratory, Technical Institute of Physics and Chemistry, Chinese Academy of Sciences (CAS)
Zhongguancun Beiyitiao 2, Haidianqu
Beijing, China, 100080

ABSTRACT
In this article, we present several approaches to replication of fine hierarchically porous structures of biomaterials and in situ synthesis of metal nanoparticles in these ceramic replicas. By using silk fiber, human air and *Pueraria Lobata* as biotemplate, we succeeded in copying their fine hierarchically porous structures via the surface sol-gel process and a newly developed immersion-fuming-calcination process. Experimental results showed that the obtained ceramic replicas had inherited the fine hierarchically porous structures of the corresponding biotemplates. These hierarchically porous ceramics were further used as nanoreactor for in situ synthesis of metal nanoparticles. The thermal properties and high-temperature stability of the composites were studied. The results indicated that the in situ synthesized metal nanoparticles were very stable, even at elevated temperatures as high as 800°C. The effects of size of metal nanoparticles and ceramic media on the stability of nanoparticles were revealed. The confinement of nanoparticles by the hierarchically porous structures largely limited the migration of nanoparticles, resulting in much improved stability of metal nanoparticles both at room temperature and at elevated temperatures.

INTRODUCTION
Through millions of years of evolution, biological materials have gained much more sophisticated structures and functions than artificial counterparts. "Learning from Nature" must be an effective way to develop novel functional materials and devices, and is currently a hot topic in materials science and many device-related researches. Recently, replication of natural materials with inorganic compositions has been attracting much attention, as they are morphologically complex with a sophisticated structure and ordering, and may act as a template for the formation of inorganic materials with potential applications in catalysis, magnetism, separation science, thermal insulation, electronics, and photonics. For example, replication of bacteria, wood cell and cellulose yielded inorganic materials of unique morphology and, in many cases, of high porosity.[1-3] Such porous materials were also used as a nanoreactor for in situ synthesis of metal nanoparticles.

Nanoparticles of metals and semiconductors are known to have unique features, which make them promising in such applications as optical, electronic, magnetic and catalytic materials and devices. The stability of nanoparticles is an important issue of particle/matrix composites, as aggregation of nanoparticles will lead to deterioration of their functionalities. It is known that nanoparticles are stabilized under ambient conditions by porous morphology of the matrix[4] and by interaction with the matrix[5]. Their high-temperature stabilities, however, are very important in regard of their practical applications such as automobile catalysis. Therefore, we also studied in the current work the thermal behavior of metal nanoparticles embedded in ceramic replicas.

EXPERIMENTAL DETAILS

Fabrication of hierarchically ordered porous titania and zirconia filaments with silk as template by the sol-gel method

Silk threads are composed of silk filaments and were used as the template for the surface sol-gel process. Ti(OnBu)$_4$ and Zr(OnBu)$_4$ in toluene (100 mM) were used as precursor solutions. In a typical procedure, silk threads were first immersed in a precursor solution for 10 min. After rinsing twice with toluene (each for 1 min), silk threads were sucked dry on a filtering unit and left in air for hydrolysis by moisture. These steps constitute 1 cycle of the surface sol-gel process, and 10 cycles were repeated. Then, the silk/metal-oxide composite threads were dried in vacuum overnight. They were finally heated at a rate of 1 K/min from room temperature to 450°C and calcined at this temperature for 4 h to remove the organic components.

Fabrication of hierarchically ordered porous titania microtubes with human hair as template by the sol-gel method

Human hair was used as template. It was washed with a domestic shampoo and thoroughly rinsed with distilled water. After drying at room temperature, it was immersed in a toluene (or ethanol) solution of Ti(OnBu)$_4$ (100 mM) for 30 min. The sample was then rinsed twice with the corresponding solvent (each for 1 min), sucked dry on a filtering unit, and left in air for hydrolysis by moisture for 30 min. These constitute one cycle of the surface sol-gel process. Different numbers of cycles were repeated. The hair/titania composite strands were dried in vacuum overnight. They were finally heated at a rate of 5 K/min from room temperature to 800°C, and calcined at this temperature for 4 h to remove the organic components.

Fabrication of hierarchically ordered porous γ-Al$_2$O$_3$ with *Pueraria lobata* as template by an immersion-fuming-calcination process.

The stem of *Pueraria lobata* (Pl) was decorticated and transversely cut into slices. To remove any possible metal ions, the Pl slices were immersed in aqueous HNO$_3$ (2.6 M) for 24 h, and rinsed repeatedly with distilled water until the pH value was nearly 7. They were dried in a vacuum at room temperature overnight. The acid-treated Pl (AT-Pl) slice was immersed in aqueous AlCl$_3$ for 70 min, placed in air for 6 h, and exposed to excessive NH$_3$ gas evaporating from ammonia solution for 3 h. After being dried at 50°C overnight, the specimen was calcined at 800°C. A white alumina slice was finally obtained.

In-situ synthesis of metal nanoparticles

A piece of the porous ceramics (TiO$_2$ and ZrO$_2$ threads, TiO$_2$ microtubes, and Al$_2$O$_3$ slices) was immersed in aqueous metal precursors (10 mM AuCl$_3$, 5 mM HAuCl$_4$, and 1 mM H$_2$PtCl$_6$, respectively) for a certain time (1 ,10, and 10 min, respectively). After being rinsed in ethanol for 30 s, it was placed in aqueous NaBH$_4$ (200 mM) for 10 min, the incorporated metal ions being reduced to metal nanoparticles. Finally, it was rinsed in pure water for 1 min and dried in vacuum overnight.

Characterization

Scanning electron microscopy (SEM) observations were carried out on a Hitachi S-5200 field emission scanning electron microscope. Transmission electron microscopy (TEM) observations were made on a JEOL JEM-2000CX transmission electron microscope at an acceleration voltage of 150 kV. Diameters of 100 or more metal nanoparticles on TEM images of each specimen were measured and analyzed using SigmaPlot 2001. X-ray diffraction (XRD) patterns were recorded on a Rigaku Dmax-2000 diffractometer using CuKα radiation.

RESULTS AND DISCUSSION
Fabrication of hierarchically ordered porous titania and zirconia filaments with silk as template by the sol-gel method[6]
Free-standing, light-yellow titania threads were fabricated using silk as template. Small shrinkage was noticed compared with the original silk thread. The width of the filament was ca. 5 μm (Figure 1a), and smaller than that (ca. 9 μm) of the original silk filament. Magnified images (Figure 1b) show that the titania filaments are in fact hierarchically porous. The large pores are

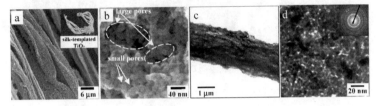

Figure 1 SEM and TEM images of the titania filaments.

ca. 100 nm in diameter. The small pores are less than 10 nm in diameter and located in the wall of the large pores. TEM images (Figure 1c) show that the filament looks spongy and has a width of ca. 2 μm. Nanopores (<10 nm) were formed in the titania filaments (Figure 1d). Thus, the highly porous character of silk was effectively inherited by the titania filament. The selected area electron diffraction (SAED) pattern (Figure 1d, inset) indicates that the filaments consist of well-crystallized anatase. Hierarchically ordered porous zirconia filaments were also fabricated similarly.

Fabrication of hierarchically ordered porous titania microtubes with human hair as template by the sol-gel method [7,8]
An array of white titania filaments (Figure 2a) was obtained when using human hair as template and 10 cycles of the surface sol–gel process were applied. Shrinkage was noticed in comparison with the original template. The width of the filaments was estimated to be ca. 40 μm from SEM images. It is smaller than that (ca. 70 μm) of the hair filaments, in agreement with the observed macroscopic shrinkage. Magnified images (Figure 2b-d) show that the filament surface is composed of titania platelets (ca. 15 μm in diameter, ca. 0.5 μm in thickness), with their planes nearly perpendicular to the filament axis. The size of the titania platelet is also smaller than that of the flat cell in the cuticle. Clearly, this unique morphology was inherited from the original hair filament, which has scales (aligned flat cells) on its surface. High magnification images (Figure 2e) show that nanopores (20-200 nm) were formed in the titania platelets. Very interestingly, the titania filaments are in fact hollow, as revealed by their cross-sectional views (Figure 2f). The inner diameter is ca. 25 μm, and the wall thickness is ca. 3 μm. Clearly, Ti(OnBu)$_4$ did not penetrate into the center of the hair filament, and the medulla was completely removed by calcination. XRD analysis (Figure 3) indicated that the TiO$_2$ microtube consists of both rutile (ca. 67 wt%) and anatase (ca. 33 wt%) crystalline phases.

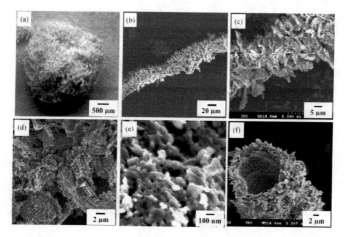

Figure 2 SEM images of the titania microtubes.

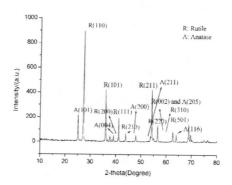

Figure 3 XRD analysis of titania microtubes.

Fabrication of hierarchically ordered porous γ-Al$_2$O$_3$ with *Pueraria lobata* as template by an immersion-fuming-calcination process.[9]

A white alumina slice was obtained using *Pueraria lobata* as template. As compared with the AT-Pl slice, the alumina slice decreased by ca. 41% in thickness and ca. 50% in diameter. Its density was estimated to be 0.43 g/cm^3, indicating its porosity. The SEM overview of the alumina slice is similar to that of the AT-Pl slice. It has four different concentric areas that were derived from the AT-Pl slice (Figure 4).

Figure 5a and b shows the magnified images of the transverse and longitudinal sections of H in Figure 4d, respectively. Clearly, H has a similar honeycomb structure to that of the pith in the AT-Pl slice (A in Figure 4b). Figure 5c and d contains magnified images of the transverse and

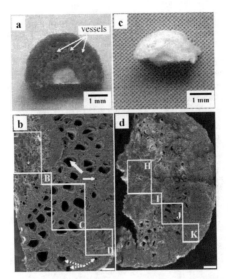

Figure 4. Digital (a and c) and SEM (b and d) images of acid treated Pl (a and b) and alumina (c and d) slices. In parts b and d, the scale bar represents 200 μm.

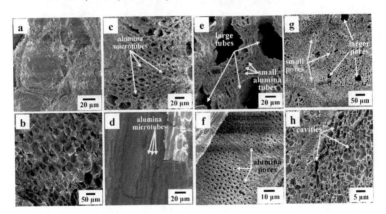

Figure 5. Magnified SEM images of transverse (a, c, e, g) andlongitudinal (b, d, f, h) sections of H, I, J, and K in Figure 4d.

longitudinal sections of I in Figure 4d. A microtube-array morphology of alumina was obtained which is the same as that of vascular bundle. Magnified images of the transverse and longitudinal sections of J in Figure 4d are shown in Figure 5e and f, respectively. Both large and small alumina microtubes were observed, and their mean diameters were estimated to be ca. 67.5 and 8.6 μm, respectively. These features were clearly inherited from the large and small vessels of C in Figure 4b. Ordered closely packed pores were observed on the wall of the large alumina microtube. They are replicas of the pores in the wall of reticulate vessels. Magnified cross-sectional image (Figure 5g and h) of K in Figure 4d shows many small and a few larger pores. They were produced by replication of parenchymatous tissue (D in Figure 4b). It was concluded that the alumina slice possesses a structure similar to that of the AT-Pl slice.

It is necessary to clarify the crystalline phases of the as-prepared alumina. As shown in its XRD pattern (Figure 6), diffraction peaks appear at 2θ = 66.76°, 45.48°, 19.08°, 36.62° and 33.12° and can be assigned to the (440), (400), (111), (311) and (220) planes of γ-alumina (JCPDS Card No. 48-367).

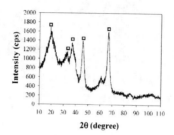

Figure 6 X-ray diffraction pattern of alumina slice.

In situ synthesis of metal nanoparticles and high-temperature stability of the composites

The nanopores in the inorganic replicas (titania and zirconia streads, titania microtubes, aluminia slices) can be used as nanoreactor for in-situ synthesis of metal nanoparticles. For example, gold nanoparticles were fabricated in both titania and zirconia streads. Their thermal stabilities were also studied.

The as-synthesized Au nanoparticles in the titania filament are small, and their mean diameter and standard deviation are estimated to be 3.9 nm (d) and 1.2 nm (σ), respectively (Figure 7a). After the composite was annealed at 500°C for 5 h, the nanoparticle size increased to 6.2 nm, the size distribution remained almost unchanged (1.3 nm (σ)) (Figure 7b). Further annealing of the composite at 800°C for 5 h resulted in a significant increase in particle size and size distribution, the mean diameter and standard deviation being 40.4 and 14.4 nm, respectively (Figure 7c). Figure 7d shows their temperature dependence. The inset in Figure 7d is a schematic illustration of the annealing process, which represents the size increase of Au nanoparticles in titania filaments. It is clear that significant increases in particle size and size distribution occur only at temperatures over 500°C.

In situ synthesis of Au nanoparticles in the zirconia filaments was conducted under identical conditions. The as-prepared Au nanoparticle has almost the same mean diameter of 4.1 nm and standard deviation of 1.0 nm (Figure 8a). After annealing at 500°C for 5 h, the size and its distribution increased only slightly (d = 5.0 nm and σ = 1.4 nm) (Figure 8b). Further annealing at 800°C resulted in larger increases in the size and size distribution (d = 10.9 nm and σ = 4.3 nm) (Figure 8c). However, the size increase at 800 °C is much smaller than that observed in the titania matrix. In other words, the Au nanoparticles embedded in amorphous zirconia are much

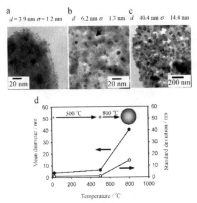

Figure 7. TEM images of Au nanoparticles in porous titania filaments (a) as-prepared, (b) after annealing at 500°C for 5 h (the arrow pointing to a small pore), (c) after additional annealing at 800°C for 5 h, and (d) temperature dependence of particle mean diameter (b) and standard deviation (O). Insert in (d): schematic illustration of annealing in which the diameters of the spheres are in the ratios of particle mean diameters.

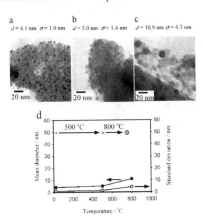

Figure 8. TEM images of Au nanoparticles in porous zirconia filaments (a) as-prepared, (b) after annealing at 500°C for 5 h, (c) after additional annealing at 800°C for 5 h, and (d) temperature dependence of particle mean diameter (b) and standard deviation (O). Insert in (d): schematic illustration of annealing in which the diameters of the spheres are in the ratios of particle mean diameters.

more stable than those in crystalline titania at 800°C. This is not surprising, if we consider that the structure of amorphous zirconia does not change much at 800°C because this temperature is far below its melting point, 2710°C. In contrast, crystal growth proceeds at 800°C for titania, and it may cause exclusion of molten Au nanoparticles, leading to their fusion to much larger particles. Figure 8d shows the temperature dependence of mean diameter and standard deviation of Au nanoparticles in zirconia. Clearly, the temperature dependence in amorphous zirconia is less significant compared with that observed in crystalline titania.

Pt nanoparticles were in situ synthesized in the γ-alumina replica and then annealed at different temperatures. The mean diameter and standard deviation of Pt nanocrystals prepared at ambient temperature were estimated to be 3.24 and 0.93 nm, respectively. The mean diameter and standard deviation from room temperature to 400°C had negligible increase. In the range 400-600°C, the mean diameter and standard deviation increased slightly faster. In this range, smaller Pt nanoparticles might be in the quasi-liquid state and begin to transfer and grow. Although the mean diameter and standard deviation increased evidently faster in the range 600-800°C than in the range 400-600°C, the increase was not significant. It is possible that Pt nanoparticles were located within the mesopores formed by alumina nanocrystals. The confining effect of alumina mesopores on Pt nanoparticles may make them less easy to transfer and grow. These results indicate that the γ-alumina replica effectively immobilized and isolated Pt nanoparticles.

The stability of metal nanoparticles at high temperatures is attributed to the porous morphology of the ceramic matrix and to the strong bonding interaction between the surface atom of the nanoparticle and the surrounding oxygen linkage of the matrix. A second important factor is melting points of metal nanoparticles. The melting point of metals is size-dependent at the nanometer scale and becomes lower with a decrease in particle size. As described above, melting of nanoparticles will promote their fusion to induce large increases in particle size and its distribution. However, the fusion may be retarded if individual nanoparticles are physically isolated from each other. Thus, the physical form of the ceramic matrix should be considered as a third factor that affects thermal stability of metal nanoparticles.

CONCLUSIONS

In conclusion, we explored two approaches to replication of fine hierarchically porous structures of biomaterials, and metal nanoparticles were in situ synthesized in these ceramic replicas. By using silk fiber and human air as template, titania and zirconia streads and titania microtubes were readily fabricated via the surface sol-gel process. By a newly developed immersion-fuming-calcination process, hierarchically ordered porous γ-Al_2O_3 was obtained using *Pueraria Lobata* as template. The obtained ceramic replicas inherited the fine hierarchically porous structures of the corresponding biotemplates. These hierarchically porous ceramics were further used as nanoreactor for in situ synthesis of metal nanoparticles. The in situ synthesized metal nanoparticles were very stable, even at elevated temperatures as high as 800°C. The confinement of nanoparticles by the hierarchically porous structures largely limited the migration of nanoparticles, resulting in much improved stability of metal nanoparticles both at room temperature and at elevated temperatures.

ACKNOWLEDGMENTS

This work was supported by "Hundred Talents Program" of CAS, the National Natural Science Foundation of China (Grant No. 20471065), and the President Fund of CAS.

REFERENCES
[1] S. A. Davis, S. L. Burkett, N. H. Mendelson, and S. Mann, *Nature*, **385**, 420-23 (1997).
[2] T. Ota, M. Imaeda, H. Takase, M. Kobayashi, N. Kinoshita, T. Hirashita, H. Miyazaki, and Y. Hikichi, *J. Am. Ceram. Soc.*, **83**, 1521-23(2000).

[3] J. Hun and T. Kunitake, *J. Am. Chem. Soc.*, **125**, 11834-40 (2003).
[4] He, J.; Kunitake, T.; Nakao, A. *Chem. Mater.*, **15**, 4401-06 (2003).
[5] He, J.; Ichinose, I.; Kunitake, T.; Nakao, A.; Shiraishi, Y.;Toshima, N. *J. Am. Chem. Soc.*, **125**, 11034-40 (2003).
[6] J. He and T. Kunitake, *Chem. Mater.*, **16**, 2656-61 (2004).
[7] S. Liu and J. He, *J. Am. Ceram. Soc.*, **88**, 3513-14 (2005).
[8] S. Liu and J. He, *J. Inorg. Mater.*, **21**, 1313-18 (2006)
[9] C. Li and J. He, *Langmuir*, **22**, 2827-31 (2006).

ELECTROSPINNING OF ALUMINA NANOFIBERS

Karin Lindqvist, Elis Carlström
IVF Industrial Research and Development
Swedish Ceramic Institute
Argongatan 30
SE-431 53 Mölndal, Sweden

Anna Nelvig, Bengt Hagström
IFP Research
Argongatan 30
SE-431 53 Mölndal, Sweden

ABSTRACT

A straightforward manufacturing route of alumina nanofibers was established by the electrospinning of a commercial alumina sol mixed with polyethylene oxide. The molecular weight and concentration of the polymer were varied as well as the spin parameters in order to find suitable fabrication settings. The results showed that the polymer content should be at least 27 vol% in the green fibre in order to avoid brittle fibres and the molecular weight of PEO should be at least 400 000 g/mol. The shear and elongation rheological properties of the suspensions were measured. A contraction flow method for measuring the elongational viscosity was evaluated and found to be suitable for highly shear thinning systems with high viscosity. The fibre diameters of calcined samples were in the range 400-700 nm. The electrospinning should be performed in a dry environment facilitating better control of the fibre deposition.

INTRODUCTION

Electrospinning is a convenient and simple process to obtain ultra thin ceramic fibres. In its simplest form a ceramic precursor suspension is placed in a syringe and subjected to a high electric potential visa-vi a grounded collector electrode. A schematic drawing of the process is shown in figure 1. When a voltage is applied to the syringe the pending spherical drop at the needle tip is changed into a conical shape (Taylor cone) and, upon a further increase in applied voltage (producing an increased charge density in the suspension), the electrostatic repulsive forces will overcome the surface tension of the suspension and a fine charged jet is ejected from the Taylor cone towards the grounded collector. During its flight the jet is highly elongated and the suspension medium is evaporating leaving a fibrous non-woven mat on the collector. The "green" fibres so produced are calcined into ceramic fibres at elevated temperature. Powder dispersions, alkoxide solutions or colloidal sols can be mixed with polymers to create the ceramic precursor suspension. During the last few years a number of different ceramics have been spun into sub-micron fibres by the electrospinning technique[1].

The rheological properties of the precursor suspension turn out to be important for the electrospinning process and for the formation of uniform fibres. Since the electrostatic forces available are comparatively weak the counter acting viscous forces have to be limited in order to create thin fibres. This will set an upper bound on viscosity. At lower viscosities surface tension will become a dominant factor and capillary break up of the liquid thread into droplets (Rayleigh instability) may produce electro spraying instead of electrospinning. To this end it is common practise to use a soluble polymer in the precursor suspension to regulate its viscoelastic properties. Several authors [2-4] have emphasised the importance of the elongational viscosity for the formation of uniform fibres. It was shown that the Rayleigh instability could be completely suppressed if a critical value of the elastic stress in the jet was exceeded. Yu [4] used an extensional rheometer measuring the filament break-up of the sample. Stading [5] developed a measurement system for shear thinning fluids that measured the elongational viscosity in contraction flow. This method gives the possibility to control the strain rate during the measurement, which is an advantage compared to measurements based on filament break-up.

The parameters controlling the final diameter of electrospun fibres have been the subject of extensive research. Sigmund [1] modified the theory for prediction of the final jet radius to account for higher bulk electrical conductivity in ceramic-polymer suspensions. According to the theory the final green fibre diameter is dependent on the surface tension, the volume flow rate of the suspension, the volume current through the jet, the electrical conductivity and the electric field strength.

A few papers report on the fabrication of electrospun aluminium oxide fibers. Azad [6] made transparent alumina fibers from aluminium pentadionate mixed with PVP in an acetone-ethanol solution. Larsen et al. [7] reported on the fabrication of alumina fibers from aged sol prepared from aluminium di-sec-butoxide ethylacetoacetat.

The objective of this study was to establish a robust straightforward process for electrospinning of alumina fibers with controlled fibre morphology. There are commercially available ceramic sols commonly used as high refractory binders. These materials have high purity, small particle size, are delivered as stable colloidal water based dispersions and could therefore be suitable as raw material for electrospinning. No papers have to our knowledge been published on manufacturing of ceramic nanofibers from commercially available sols. Several water-soluble polymers could be candidates for facilitating the formation of ceramic fibers. Polyethylene oxide (PEO) is commonly used in electrospinning and was chosen in this study. The molecular weight and concentration of the polymer have been varied as well as the spin parameters in order to find suitable fabrication settings. The shear and elongation rheological properties of the suspensions have been measured. A contraction flow method for elongational viscosity was evaluated to see if this method was suitable for electrospinning suspensions. The fibre diameters were measured on calcined samples.

EXPERIMENTS

A commercially available colloidal alumina sol, Nyacol AL20 (Nyacol Nano Technologies Inc., USA) with a primary particle size of 50 nm was used as the ceramic precursor. Polyethylene oxide with molecular weights (Mw) of 100,000 (100k), 400,000 (400k) and 900,000 (900k) (Sigma Aldrich, Sweden) were dissolved in the sol by magnetic stirring so that a suspension of polymer and sol was obtained. The conductivity (CDM210, Radiometer Analytical, France) was measured on the alumina sol as received and with the polymer dissolved. The suspensions were rheologically characterized by measuring the steady state shear viscosity and the phase shift in oscillatory shearing in a cup and bob (CC25) measurement system (StressTech, Rheologica, Sweden). The elongational viscosity was estimated through the measurement of the pressure drop over a specially designed contraction flow nozzle (Reologen i Lund, Sweden) that produces a constant elongational strain rate along its centre line, see figure 2. The measurement of elongational viscosity requires a rather strong shear thinning behaviour of the fluid in order to subtract the contribution from shear flow components to the measured pressure drop.

Electrospinning was performed in the experimental set up schematically shown in figure 1. Initial experiments were performed in ambient atmosphere with the electrospinning parameters set to 20 kV (high voltage power supply ES50P from Gamma High Voltage, Ormon Beach, FL, USA), a syringe-collector distance of 20 cm; a syringe of 0.8 mm and a flow rate of 0.056 ml/min. A grounded platinum foil was used as collector electrode. Electrospinning for fibre diameter measurements were performed in a Plexiglas box under flowing nitrogen gas with parameters set to obtain a single well defined Taylor cone at the tip of the syringe needle. The parameters for each system are shown in table 1.

The fibres were calcined at 700°C in air. XRD was used to identify the alumina phases obtained after calcinations. The specific surface area was measured (Gemini 2300, Micromeretics) on green and calcined fibres. Samples for fibre diameter measurements were prepared by sedimentation of calcined and crushed fibres on a sample holder. The fibres were manually measured from SEM micrographs.

 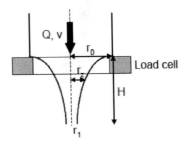

Figure 1. Schematic of the electrospinning process

Figure 2. Measuring system for contraction flow measurement.

Table 1. Parameters for electrospinning of fibers for diameter measurements.

Sample	Composition	Voltage (kV)	Collector distance (cm)	Field strength (kV/cm)	Flow rate (ml/min)
A	3wt% 400k	10	15	0.67	0.05
B	6wt% 400k	20	15	1.33	0.05
C	3wt% 900k	10	10	1	0.005
D	3wt% 900k	10	10	1	0.01

RESULTS
The results from the shear viscosity measurements and frequency sweep measurements are shown in figures 3 and 4.

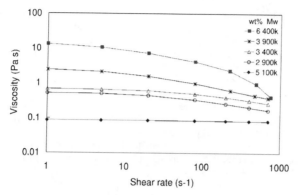

Figure 3. Shear viscosity vs. shear rate for polyethylene oxide dissolved in alumina sol.

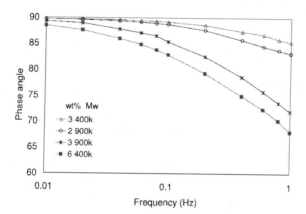

Figure 4. Phase angle vs frequency for PEO dissolved in alumina sol

Figure 3 shows that the suspension with 5 wt% PEO 100k was Newtonian and had a low shear viscosity. No fibres were obtained with this sample but spherical particles from electrospraying, see figure 5. Severe dripping during handling also made this suspension unsuitable for electrospinning and no further analyses were made. The suspensions with PEO 400k and PEO 900k had higher shear viscosities and were shear thinning. Suspensions with higher concentrations of PEO 400k or 900k could not be dissolved properly in the sol due to very high viscosities.

Oscillatory measurements in shear were performed to obtain the viscoelastic properties and the results are shown in figure 4. A phase shift of 90 degrees between strain and stress means that the material is completely viscous. The lower the phase shifts the more elastic is the suspension. Although the shear viscosity is moderate for the sample with 3wt% PEO 900k the elasticity is high at high frequencies indicating the importance of molecular weight for this property.

Preliminary electrospinning experiments were performed in ambient environment without controlling humidity. The fibre formation process was erratic and fibres were deposited not only on the collector electrode but also on surrounding surfaces. The fibres were in fact "standing up" forming a 3-dimensional cotton like morphology in the applied electric field and fell back (more or less) on the collector when the voltage was turned off. Turning on the voltage again resulted in the "erection" of the fibres. It seems as the fibres were polarised in the field. The formation of 3-dimensional cotton like structure has so far not been recorded when pure polymer solutions are electrospun to fibres. However, Azad [8] obtained this type of behaviour when processing zirconia and ceria fibres in ambient atmosphere.

Samples with 2wt% polymer in the alumina sol had a lot of broken fibres when studied in SEM, see figure 6. The sample was very brittle, which is believed to be due to the low polymer content (20 vol% in the green fibre). An addition of 3 wt% polymer resulted in more continuous fibres and no broken fibres were found in samples with 6wt % polymer added, see figure 7.

Figure 5. 5 wt% PEO 100k. Spherical particles from electrospraying, green sample.

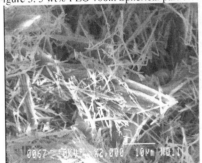

Figure 6. 2wt% PEO 900k.   Green broken   Fig 7. 6wt% PEO 400k green fibres.
fibres due to low polymer content.

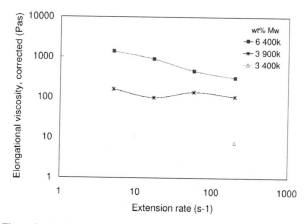

Figure 8. Elongational viscosity vs. extension rate for PEO alumina sol suspensions.

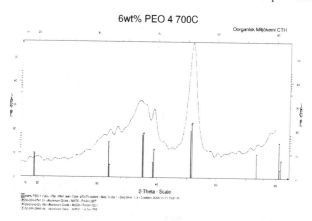

Figure 9. Results from XRD measurements on calcined fibres of 6wt%PEO 400k pressed into a thin disc shape.

The elongational viscosity and fibre diameters were measured on the systems shown in table 1. Figure 8 shows the results from the measurements of elongational viscosity with the results corrected for the shear viscosity that has an influence on the registered force during the measurements. Only one extension rate could be accurately measured for the system with 3 wt% PEO 400k due to the low shear viscosity and the limited shear thinning of that system. There is a much larger difference between the behaviour of the systems in extension than in shear. For a Newtonian fluid the elongational viscosity is three times (Trouton's ratio) the shear viscosity. In this case, when the molecular weight of PEO is sufficiently high, the elongational viscosity is 50-100 times larger than the shear viscosity and it can be assumed that the suspensions are strain hardening in elongation. Such behaviour will stabilise the filament and prevent capillary break up.

Very little information about the actual elongational strain rates during electrospinning is available in the literature. Yu measured the jet thinning of aqueous PEO solutions from the Taylor cone up to the point where the whipping instability sets in. Based on the flow rate and the diameter progression along the spin line the strain rate can be measured in this region. Data from Yu indicate strain rates in order of magnitude around 10 s^{-1}. Unfortunately, the strain rates in the whipping instability region are not known since it is difficult to assess the point along the fibre where stretching ceases. . The elongational rheometer used in this study cover a wide range of strain rates around 10 s^{-1} so it seems reasonable to assume that it provides viscosity data in the relevant strain rate regime.

Figure 9 shows that the main phase obtained when calcining to 700°C was γ -alumina. There is no amorphous phase in the sample but the broad peaks show that the grain size is very small, about 30 nm.

The fibre diameters of calcined samples obtained from measurements on SEM micrographs are given in Table 2. A frequency diagram is shown in figure 10. All of the systems had uniform fibres without branches when the electrospinning parameters were set to give a single Taylor cone, which is required to get proper whipping of the jet. Comparing system C and D, having the same composition and electric field strength shows that an increased flow rate resulted in larger fibre diameter. Figure 11 shows a SEM micrograph of system D with 3 wt% PEO 900k.

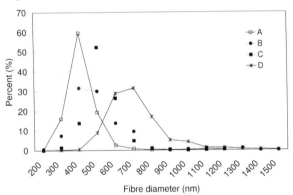

Figure 10. Calcined fibres diameter frequency diagram

Figure 11. 3 wt% PEO 900k calcined fibres electrospun in Plexiglas box under flowing nitrogen.

Table 2. Results from measurements of conductivity, fibre diameter and specific surface area.

Sample	Composition	Conductivity (mS/cm)	Average fibre diameter (nm)	Standard deviation (nm)	Confidence interval (95%)	Specific surface area (m^2/g)
	Alumina sol	2.2	-			-
	3wt% 100k	2.0	-			-
A	3wt% 400k	2.0	421	112	± 11	166.9
B	6wt% 400k	1.8	557	330	± 47	165.6
						33.3 (green)
C	3wt% 900k	1.9	526	97	± 8	-
D	3wt% 900k	1.9	710	160	± 16	158.5

The value of the conductivity was independent of the molecular weight of the PEO and seems to be influenced mainly by the concentration of the sol in the sample. The specific surface area of the green sample of 6wt% PEO400k was only 33.3 m^2/g while the calcined samples had much higher values. The large specific surface area of the calcined samples is mainly due to the porous crystal structure of the γ - alumina. The surface of the alumina is "hidden" by the polymer before calcination of the sample.

DISCUSSION

The choice of a commercial alumina sol as a source for the ceramic phase of electrospun nanofibres resulted in a straightforward preparation of the spinning suspension. The choice of polymer, molecular weight and concentration is crucial for the formation of uniform fibres. The initial electrospinning in this study showed that suspensions with PEO100k did not form continuous fibres but resulted in spherical particles due to the Rayleigh instability. Higher molecular weight or concentration of the polymer resulted in uniform fibres. However, too low polymer content gave very brittle fibres and green fibres broke during handling. It can therefore be concluded that a minimum amount of polymer, in this study about 27 vol%, is necessary to obtain continuous and usable fibres.

The fixed initial spinning conditions often resulted in several simultaneous Taylor cones at the tip of the syringe needle. Spinning was also discontinuous in the sense that the cones appeared and disappeared in a stochastic fashion and drops were accumulated and released periodically at the needle tip. It was soon realised that in order to have one single active Taylor cone and a continuous spinning without dripping it was necessary to tune the spinning conditions for each suspension. To avoid multiple jets and dripping, as compared to the initial settings, it was necessary to decrease the flow rate from the syringe pump and/or increase the field strength by increasing the voltage or decreasing the distance to the collector. The parameter combinations in Table 1 resulted in a single well defined Taylor cone and continuous spinning without dripping. There is a processing window for each suspension allowing a stable and well defined spinning.

It was noted that the ambient humidity greatly affected the deposition of fibres. By using nitrogen blanket the relative humidity was brought close to zero and the fibres were then directed and deposited on the collector. Humidity may affect the surface resistivity of surrounding objects made from otherwise insulating materials like wood, paper and plastics via adsorption and it can be speculated that a low humidity will concentrate the electrical field lines from the charged syringe towards the grounded collector plate. However, the deposition was still 3-dimensional in the sense that a "cotton like" structure was built up from the collector surface, see figure 12.

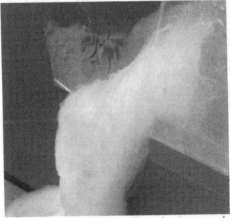

Figure 12. Morphological feature of green alumina fibres electrospun under flowing nitrogen.

This type of highly porous structure may be interesting for certain applications like high temperature filtration. This type of structure is not seen when pure polymer solutions are electrospun. In this case a 2-dimensional non-woven mat is formed on the collector electrode. This type of 2-dimensional structure was also recorded when alumina particles were dispersed in aqueous PEO solutions and electrospun.

According to the theory [1] of predicting the green fibre diameter an increased flow rate and /or a decreased field will increase the fibre diameter. The bulk conductivity and the surface tension of the suspension will also have an influence on the fibre diameter. In the systems studied the conductivity is almost constant and the differences in surface tension will also be very small. The fibre diameters as influenced by the electrospinning parameters were not studied in a systematic manner. However, some observations regarding the possible effect of elongational viscosity can still be made. Comparing system A (3wt% PEO 400k) and B (6wt% PEO 400k) with the same flow rate but B having twice the field strength one would expect B to show a smaller diameter than A if the elongational viscosity had no effect. In fact, B shows thicker fibres. The higher elongational viscosity due to the higher polymer concentration in B might explain this. A similar argument is found by comparing A and C (3wt%PEO 900k). C having higher field strength and significantly lower flow rate than A, both should work in the direction of thinner fibres, still C shows thicker fibres. It should be noted that the fibre diameters were measured on calcined samples obtained by sedimentation on a sample holder. This sample preparation is not possible to use for green fibres that would dissolve during the sedimentation. However, the shrinkage of the calcined fibres will be the same for all suspensions containing 3wt%PEO regardless of the molecular weight. For the system containing 6wt% the shrinkage will be larger; that is the green fibres were coarser than measured on the calcined samples. This further stresses the influence of the elongational viscosity as stated above.

The rheological measurements showed that there is a much larger difference in the extensional behaviour between the systems than in the shear behaviour. Contraction flow measurements are suitable for highly shear thinning systems with a rather high viscosity. This method would have been more suitable if the polymer suspension contained polymer of higher molecular weights or higher concentrations. Extensional rheology measurements based on filament break up is probably more suitable and also possible to use on solutions with polymers of low molecular weight and concentration.

CONCLUSIONS

We have shown that a straightforward manufacturing of alumina nanofibers is possible by the electrospinning of a commercial alumina sol mixed with polyethylene oxide. The polymer content should be at least 27 vol% and then have a molecular weight of at least 400 000. The concentration range should be 3-6 wt% of polymer in the sol. The electrospinning should be performed in a dry environment facilitating better control of the fibre deposition. The manufactured fibres had diameters in the range 400-700 nm and further optimisation can reduce these values.

REFERENCES

1. Sigmund, W., et al., Processing and Structure Relationships in Electrospinning of Ceramic Fiber Systems. *J. Am. Ceram. Soc.*, **89**(2), 395-407.(2006).
2. Daga, V., Helgeson, M., and Wagner, N., Electrospinning of Neat and Laponite-Filled Aqueous Poly(ethylene oxide) Solutions. *Journal of Polymer Science: Part B: Polymer Physics*, **44**, 1608-1617.(2006).
3. Shenoy, S.L., et al., Role of chain entanglements on fiber formation during electrospinning of polymer solutions: good solvent, non-specific polymer-polymer interaction limit., *Polymer*, **46**, 3372-3384.(2005).
4. Yu, J.H., Fridrikh, S.V., and Rutledge, G.C., The role of elasticity in the formation of electrospun fibers. *Polymer*, **47**, 4789-4797.(2006).
5. Stading, M. and Bohlin, L. Contraction flow for characterization of extensional properties of semi-solid materials. in *XIVth International Congress on Rheology*. 2004. Seoul, Korea: The Korean Society of Rheology.

6.   Azad, A.-M., Fabrication of transparent alumina (Al2O3) nanofibers by electrospinning. *Materials Science and Engineering: A*, **435-436**, 468-473.(2006).
7.   Larsen, G., et al., A method for making Inorganic and Hybrid (Organic/Inorganic) Fibers and vesicles with Diameters in the Submicrometer and Micrometer range via Sol-gel Chemistry and Electrically Forced Liquid Jets. *J. Am. Chem: Soc.*, **125**(1154-1155).(2003).
8.   Azad, A.-M., Matthews, T., and Swary, J., Processing and characterization of electrospun Y_2O_3-stabilized ZrO_2 (YSZ) and Gd_2O_3 -doped CeO_2 (GDC) nanofibers. *Materials Science and Engineering: B*, **123**, 252-258.(2005).

CARBON NANOTUBE (CNT) AND CARBON FIBER REINFORCED HIGH TOUGHNESS REACTION BONDED COMPOSITES

P. G. Karandikar, G. Evans, and M. K. Aghajanian
M Cubed Technologies, Inc.
1 Tralee Industrial Park
Newark, DE 19711

ABSTRACT
Reaction bonded SiC and B$_4$C offer low density (light weight), high hardness, high stiffness, high thermal conductivity, low CTE, excellent ballistic resistance, complex shape producibility and high volume producibility – properties needed in many armor, thermal management, semi conductor capital equipment, and aerospace mirrors and structures markets. However, similar to other ceramics, their toughness and strength are lower compared to incumbent metallic materials. Here we demonstrate the use of carbon fiber and carbon nanotube (CNT) reinforcements for increasing toughness of reaction bonded ceramics. The reaction bonding process depends on the reaction between carbon and molten silicon to achieve infiltration of particulate or fibrous preforms (e.g. SiC, B$_4$C, etc.). Thus, if unprotected, carbon fibers and CNTs will convert to SiC in this process. In this work, innovative processing was conducted to successfully incorporate both carbon fibers and carbon nanotubes in reaction bonded materials. C$_f$/SiC composites were obtained with quasi-isotropic low CTE (< 1 ppm/K) and high fracture toughness (6-10 MPa m$^{1/2}$). Fracture toughness of reaction bonded SiC was increased from 4 to 7 MPa m$^{1/2}$ (a 73% increase) using CNTs.

INTRODUCTION
Reaction bonded SiC and B$_4$C materials[1-11] offer high specific stiffness (stiffness/density - E/ρ) and thermal stability (thermal conductivity / coefficient of thermal expansion - k/CTE), and are used for applications such as lithography equipment, optical structures, mirrors, thermal management and aerospace components. Properties of several candidate materials that were historically considered for these applications are compared with those of M Cubed materials in Table 1.

In some aerospace applications however, high toughness and strength are also required. Beryllium (Be) offers the best combination of specific stiffness, high strength and high toughness (8-10 MPa m$^{1/2}$), and is therefore extensively used in aerospace structural components. However, Be has many limitations such as high cost, low thermal stability and above all, health hazards associated with Be dust. Reaction bonded SiC and B$_4$C can compete with Be in terms of specific stiffness and have better thermal stability than Be. However, they suffer from the limitation of low (4 MPa m$^{1/2}$) toughness.

THE REACTION BONDING PROCESS
Reaction bonded SiC was first developed in the 1940's[1-5]. Other terms for the process include 'reaction sintered' and 'self bonded". Figure 1 shows a schematic of this process. In this process, the preform containing the reinforcement and a carbon precursor or binder is "carbonized" in an inert atmosphere above 600°C to convert the precursor to carbon. Next, the preform is placed in contact with Si metal or alloys of Si in an inert or vacuum atmosphere and

heated to above the melting point of the alloy. Due to the spontaneous wetting and reaction between carbon and molten Si, the preform is infiltrated completely. The carbon in the preform reacts with the Si forming SiC and in the process bonds the reinforcement together. Some residual Si remains.

Table 1. Summary of properties of selected materials for aerospace applications.

Material	ρ (g/cc)	α (ppm/K)	k (w/mK)	E (GPa)	K_{IC} (MPa/m$^{1/2}$)	UBS or (UTS) (MPa)	E/ρ	k/α	Strength/ ρ
Al	2.7	27	237	70	20	270	26	8.8	100
Be	1.85	11.4	150	300	10	(324)	162	13.1	(175)
Si	2.33	2.6	150	113	1-2	50-80	48	58	22
CVD SiC	3.22	2.4	175	364	4	450	123	73	150
ULE	2.2	0.03	1.3	73	1-2	50-90	33	43	23
Zerodur	2.55	0.05	1.6	80	1-2	50-90	36	39	20
Silica	2.2	0.65	1	70	1-2	32	30	3	15
M Cubed Materials									
Al/SiC (55)	2.95	10	180	200	11	(340)	68	18	(115)
Al/SiC (70)	3.01	4.1	170	270	10	(230)	90	41	(77)
Si/SiC (80)	3.03	2.9	185	380	4	290	125	64	96
Si/B$_4$C (70)	2.57	4.0	100	382	4.8	271	149	25	105

ρ = Density, α = CTE (-50 to 100°C), k = Thermal Conductivity, E = stiffness, K_{IC} = Fracture Toughness

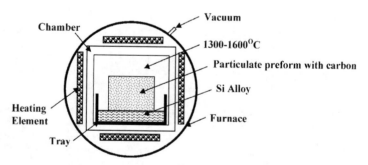

Figure 1. A schematic of the reaction bonding process

A major advantage of this process is that the volume of the reaction-formed SiC is 2.3 times larger than the volume of the reacted carbon. Thus, by infiltrating Si into preforms that contain high carbon contents, ceramic bodies rich in SiC can be produced. In addition, like water, silicon expands on freezing. As a result, unlike other materials made by casting-type processes, RB composites do not show any shrinkage porosity. The reaction bonding process has several advantages relative to traditional ceramic processes (e.g., hot pressing, sintering). First and foremost, volume change during processing is very low (generally well less than 1%), which provides very good dimensional tolerance control and eliminates the need for final machining. In addition, the process requires relatively low process temperatures and no applied pressure, which reduces capital, tooling and operating costs. Moreover, fine high surface area powders capable of being densified are not required, which reduces raw material cost.

In the earlier work[5] on reaction bonded B$_4$C it was shown that an unwanted reaction between the Si and B$_4$C phases would often occur during the infiltration process, which resulted in cracking of the parts. To prevent excessive reaction from occurring, coarse particles (e.g. 300 microns) were used. This limited the surface area for reaction, but also limited the strength of the resultant material.

M Cubed has further optimized the reaction bonding process[6-11], to produce relatively fine grained SiC and B$_4$C ceramics with favorable mechanical and ballistic properties as well as good machinability. In the specific case of reaction bonded B$_4$C, the unwanted reaction between Si and B$_4$C was suppressed by alloying the Si infiltrant with boron (B). The binder formulation was optimized to yield preforms with high strength to allow intricate green machining using CNC machines. Preform bonding technology was developed to allow fabrication of complex shapes, ribbed structures, box structures, cooling channels etc. The shrinkage from the preform stage to final infiltrated stage was reduced to less than 0.5%, allowing net-shape component fabrication with minimal finish machining (that too only for high precision components). This process technology allows fabrication of intricate large structures (500 lb, 1.5m x 0.75 m x 0.3 m), as well as mass production of smaller components (10s of thousand per month).

In spite of these refinements, one limitation of these materials remains, namely their lower fracture toughness than Be. As a result, their applications have been limited to stiffness controlled applications and they are not viable for applications needing high strength and toughness. Therefore, the present work pursued two approaches for enhancing the fracture toughness of reaction bonded materials: (1) Carbon fiber reinforcement, (2) carbon nanotube (CNT) reinforcement.

CARBON FIBER REINFORCED RB SiC

Fiber reinforcement (e.g. SiC, Al$_2$O$_3$, and carbon) has been shown to increase toughness of ceramics[12]. Out of all the fibers, carbon fibers are the most cost competitive due to their wider-scale usage in polymeric composites (e.g. graphite-epoxy). Other fibers (e. g. SiC, Al$_2$O$_3$) have proven to be cost prohibitive and their use is limited to very high temperature applications (>1000°C) where carbon fibers cannot be used due to their oxidation problem. Continuous carbon fiber based composites are typically anisotropic and laminate designs are used for making most of the components. The laminates are typically quasi-isotropic (i.e. the properties are uniform in the plane of the laminate).

As described earlier, the reaction bonding process depends on the good wetting and chemical reaction between carbon and molten silicon. This wetting and reaction lead to spontaneous infiltration of preforms containing carbon by molten silicon. For processing carbon fiber reinforced SiC composites by reaction bonding, the key challenge is prevention of attack on the carbon fibers by molten silicon and their conversion to SiC. In the past[12-13] this problem was overcome using carbon, SiC or BN coatings deposited on the fibers in a separate process step by techniques such as chemical vapor deposition (CVD) or chemical vapor infiltration (CVI). Such processes are time consuming and expensive.

M Cubed has developed[14] in-situ formed coatings that protect carbon fibers from attack by molten silicon. Figure 2a shows a micrograph of a woven C$_f$/SiC composite made by reaction bonding and using the in-situ formed protective coating. Energy dispersive analysis of X-rays (EDAX) was conducted on the fibers in this microstructure. Figure 2b shows an EDAX pattern from one such fiber. In this pattern, predominantly carbon is seen with very minor amount of Si. Thus, the fiber was not converted to Si. Typically, the EDAX pattern of SiC predominantly shows Si because in the presence of silicon, carbon X-rays get absorbed and do not reach the detector. Thus, the fact that the EDAX pattern from the fiber shows predominantly carbon is very significant in proving that the fiber was not converted to SiC. A dense carbon coating formed on the fiber surface during the initial processing steps is postulated to form a dense, protective SiC coating on the fiber surface during silicon infiltration, which prevents further attack by molten silicon and complete conversion to silicon carbide[14-16].

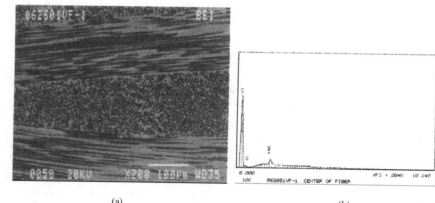

(a)   (b)

Figure 2. (a) Microstructure of C$_f$/SiC composites (b) EDAX analysis from fibers shows predominantly carbon indicating that the fibers are protected from reacting with molten silicon.

To-date several different C$_f$/SiC composites have been made with different fibers, fiber architectures and in-situ formed coatings. The focus of the present development effort was obtaining composites with quasi isotropic low CTE and high toughness for some specific aerospace applications. Typically 2 x 2 x 0.125 and 6 x 6 x 0.125 inch panels were fabricated. The C$_f$/SiC composite plates were subjected to extensive microstructural, physical, thermal and mechanical characterization. The densities of the test samples were determined using the Archimedes principle per ASTM C373. For microstructural observations, specimens were sectioned and polished and observed by optical and scanning electron microscopes (SEM). Specimen elastic moduli were measured using strain gages in a tensile test. The flexural strengths were determined using a four-point bend testing apparatus per ASTM C1161. Fracture toughness measurements were made on selected specimens by the Chevron notch method

(ASTM C1421). Strength and toughness measurements were made on at least 5 specimens of each type. CTE was measured by laser interferometry. Conventional dilatometry techniques are not sufficiently accurate for measuring CTEs of low expansion materials. Thermal conductivities were measured by the axial rod method. The measured properties of these composites are summarized in Table 2.

All three composites on which toughness measurements were made show toughness in excess of 6 MPa m$^{1/2}$. The data also shows that very low CTE ($<$ 1 ppm/K) C$_f$/SiC composites were successfully produced. It also shows that the CTE of the composite can be controlled by changing the fiber type (pitch based, PAN based, high modulus, high strength, extent of graphitization etc.) and architecture (unidirectional, cross ply, quasi-isotropic, woven etc.). The ultimate bend strength (UBS) and the modulus (E) of each composite are controlled by the fiber architecture, interface type (fiber pull out) and the extent of fiber-silicon reaction.

Table 2.   Properties of C$_f$/SiC composites made by reaction bonding for low CTE applications

Lay Up	Inter-face*	ρ (g/cc)	UBS (MPa)	K$_{IC}$ (MPa m$^{1/2}$)	E (GPa)	α (ppm/K)	K (W/mK)
8HS	A	2.39	65	--	155	0.96	68
PW	B	2.00	156	--	61	0.77	16
Q-Iso	E	2.45	109	6.0	--	1.06	91
8HS	A	2.57	184	9.4	--	1.75	143
8HS	C	2.35	126	--	--	1.84	94
8HS	B	2.49	162	--	112	-0.46	x: 114 y: 122
Uni-directional	E	2.22	312	9.7	--	-1.02	--

*The interface designations refer to various proprietary in-situ coatings.

Several components have been fabricated from these composites. Figure 3 shows a photo of a lightweight mirror made out of C$_f$/SiC. Work continues to further enhance the toughness and strength of these composites.

Figure 3. Photos of the back and the front of a lightweight C$_f$/SiC mirror.

ADVANTAGES OF CNT REINFORCEMENT AND PROCESSING CHALLENGES

Carbon nanotubes (CNTs) were first observed by Iijima in 1991. Since then, a significant effort has been spent in trying to make and characterize nanotubes[17-22]. The properties of two typical carbon fibers are compared with the properties of carbon nanotubes in Table 3. Clearly, the properties of CNTs are significantly better than those of carbon fibers. Therefore, CNTs offer a very high reinforcing potential.

Table 3. Comparison of properties of carbon nanotubes with the properties of carbon fibers.

Property	Carbon Fibers		Carbon Nanotubes
	T300 (PAN based)	P120 (pitch Based)	
Diameter (micrometer)	7	10	0.05
Density (g/cc)	1.76	2.17	~2.0
Elastic Modulus (GPa)	231	827	1000-1400
Ultimate Tensile Strength (GPa)	3.75	2.41	7-10
Thermal Conductivity (w/mK)	8	640	>2000
CTE (ppm/K)	-0.6	-1.45	-1 (isotropic)
Electrical Resistivity (micro-ohm-m)	18	2.2	<0.1

*The mechanical, thermal and electrical properties are in the axial direction.

CNTs are made by a variety of techniques such as chemical vapor deposition (CVD), laser evaporation, carbon-arc etc. CNTs are available in either single walled or multi-walled form. Figure 4 shows SEM of CVD-grown multi-walled CNTs. It can be seen from Figure 4 that the CVD-grown multi-walled CNTs (the most cost-effective form) are inter-twined like spaghetti. Thus, effective means are required to de-agglomerate the CNTs.

Processing of CNT composites presents many challenges: (1) Difficulty in uniformly dispersing , (2) Difficulty in aligning in certain directions, (3) Achieving a good CNT-matrix bond, (4) In the case of ceramic matrices – preventing reaction between the CNTs and matrix due to high processing temperature/pressure, and (5) In the case of reaction bonding, an additional challenge is the attack and conversion to SiC by molten silicon.

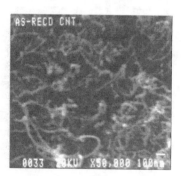

Figure 4. SEM of CVD grown multi walled carbon nanotubes (MWNTs).

FABRICATION OF RB CNT/SiC

As described earlier, M Cubed has developed a technique to protect carbon fibers from attack by molten silicon during reaction bonding[14]. This technology was extended to process CNT/SiC composites by reaction bonding[22]. In addition, process modifications were made to overcome some of the processing difficulties listed in the previous section (e.g. dispersing CNTs). This technology was applied to make several 2 x 2 x 0.25 inch sample plates for characterization.

The CNT/SiC composite plates were subjected to extensive microstructural, physical, thermal and mechanical characterization. The densities of the test samples were determined using the Archimedes principle per ASTM C373. For microstructural observations, specimens were sectioned and polished and observed by optical and scanning electron microscopes (SEM). Specimen elastic moduli were measured by an ultrasonic measurement technique. The time of flight of a shear wave and a longitudinal wave through the sample were measured to determine the Poisson's ratio of the material, as well as the elastic modulus. The flexural strengths were determined using a four-point bend testing apparatus per ASTM C1161. Fracture toughness measurements were made on selected specimens by the Chevron notch method (ASTM C1421). Strength and toughness measurements were made on at least 5 samples of each type. CTE was measured between 0 and 100°C using dilatometry. Thermal conductivity (K) was calculated from measured thermal diffusivity (D) and specific heat (Cp) data (K = Cp . ρ . D; where ρ is the density of the material). Thermal diffusivity was measured using a laser flash method (ASTM E1461-92). Specific heats were also measured by the laser flash technique. Electrical resistivity measurements were carried out using the four point probe.

Figure 5 compares the microstructure of a CNT/SiC composite with the microstructure of a standard reaction bonded SiC. Identical starting SiC powders were used for making both materials. The SiC content in the CNT/SiC is a little bit higher due to higher amount of reaction formed SiC. The property data for the two materials are compared in Table 4. The table shows that the fracture toughness of the standard reaction bonded SiC was increased by 73% using CNT reinforcement, while maintaining or improving all other properties. Also, a 78% reduction was obtained in electrical resistivity by using CNTs. This is another indication that CNTs still exist in the final composite and are not converted to SiC. As listed in Table 3, CNTs have a very low electrical resistivity. On the other hand, SiC has high electrical resistivity.

Figures 6 and 7 show high resolution SEM micrographs of the fracture surface of two CNT/SiC test specimens. Here individual CNTs can be seen pulled out of Si/SiC. These observations unequivocally demonstrated that the CNTs were successfully protected from complete reaction with molten silicon. Similar to the case of carbon fibers, it is postulated that CNTs are protected by an in-situ formed SiC coating. Based on these microstructural observations, the higher toughness in CNT/SiC composite compared to the baseline SiC can be attributed to one or more of the following mechanisms: crack deflection by CNTs, crack bridging by CNTs, and CNT pullout.

A CNT/SiC demonstration component was successfully fabricated. A photo of this component is shown in Figure 8. The high fracture toughness obtained in the CNT/SiC composite while maintaining most of the other properties of the baseline RB SiC, will allow this material to start competing with Be for many aerospace applications. Work is underway to further enhance the strength and fracture toughness of this material.

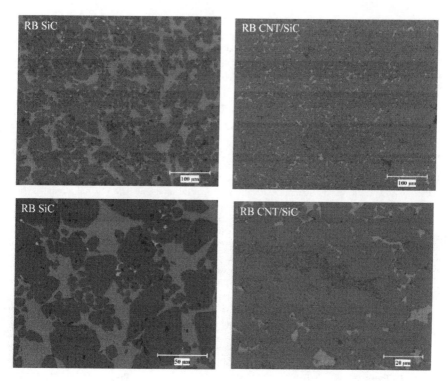

Figure 5. Comparison of microstructures of the RB SiC and RB CNT/SiC.

Table 4. Comparison of properties of RB CNT/SiC and RB SiC.

Property	Standard RB SiC	CNT/SiC	% Increase (Decrease)
Density (g/cc)	3.02	3.06	1.3
Stiffness (GPa)	359	374	4
Bend Strength (MPa)	288	285	(1)
Toughness (MPa m $^{1/2}$)	4	6.9	73
Electrical Resistivity (Ohm-cm)	1.097	0.245	(78)
CTE 25-100C (ppm/K)	2.9	2.72	(6)
Thermal Conductivity (w/mK)	188	185	(1.5)

Figure 6. SEM of the fracture surface of RB CNT/SiC example sample 1. The higher magnification picture clearly shows individual CNTs pulled out.

Figure 7. SEM of the fracture surface of RB CNT/SiC example sample 2. The higher magnification picture clearly shows individual CNTs pulled out.

CONCLUSIONS

Reaction bonded composites offer very high mechanical and thermal stability. However, like other ceramics, their toughness is lower than metallic materials such as Al and Be. Carbon fibers and CNTs can enhance toughness of reaction bonded ceramics. However, both these reinforcements can be attacked by molten silicon during the reaction bonding process. In this work, C_f/SiC composites were successfully fabricated by reaction bonding by using innovative in-situ formed protective coatings. C_f/SiC composites with toughness of 6 to 9 MPa m$^{1/2}$ and quasi-isotropic low CTE (<1 ppm/K) were produced. Similarly, CNTs were successfully dispersed and protected during RB SiC composite fabrication. Fracture toughness of RB SiC was increased by 73% using CNT reinforcement.

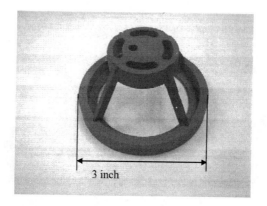

3 inch

Figure 8. A photo of a CNT/SiC demonstration component.

ACKNOWLEDGEMENTS

This work was supported by the following US DoD SBIR contracts: F29601-03-C-0012, FA9453-04-M-0255, and HQ0006-06-C-7443.

REFERENCES

[1] K.M. Taylor, "Cold Molded Dense Silicon Carbide Articles and Methods of Making the Same," U.S. Pat. No. 3 205 043, Sept. 7, 1965.

[2] P.P. Popper, "Production of Dense Bodies of Silicon Carbide," U.S. Pat. No. 3 275 722, (1966).

[3] C.W. Forrest, "Manufacture of Dense Bodies of Silicon Carbide," U.S. Patent No. 3 495 939, (1970).

[4] R. Morrell, *Handbook of Properties of Technical and Engineering Ceramics,* HMSO Publications, London, U.K., 1985.

[5] K.M. Taylor and R.J. Palicka, "Dense Carbide Composite for Armor and Abrasives," U.S. Pat. No. 3 765 300, (1973).

[6] M. Aghajanian, B. Morgan, J. Singh, J. Mears and B. Wolffe, "A new family of reaction bonded ceramics for armor applications," in Ceramic Armor Materials by Design, Ceramic Transactions, Vol. 134., J. W. McCauley et al editors, (2002) 527-540.

[7] P. Karandikar, M. Aghajanian, and B. Morgan, "Complex, net-shape ceramic composite components for structural, lithography, mirror and armor applications," Ceramic Engineering and Science Proceedings, Vol. 24 [4], (2003) 561-566.

[8] M. Waggoner, B. Rossing, M. Richmond, M. Aghajanian, and A. McCormick, "Silicon carbide composites and methods for making same," US Patent No. 6,503,572 (2003)

[9] P. G. Karandikar and M. K. Aghajanian, "Ultrasonic NDE of reaction bonded ceramics," Ceramic Transactions 175, Advances in CMC-XI, (2006) 49-61.

[10] M. Waggoner, B. Rossing, M. Richmond, M. Aghajanian, and A. McCormick, "Silicon carbide composite bodies and methods for making same," US Patent No. 6,919,127 (2005).

[11] P. G. Karandikar, M.K. Aghajanian, D. Agrawal, and J. Chang, "Microwave Assisted (MASS) processing of metal-ceramic and reaction bonded composites", Ceramic Engineering and Science Proceedings Vol. 27 [2] (2007) 435-446.

[12] P. G. Karandikar and Tsu-Wei Chou, "Chapter 10: Structural Properties of Ceramic Matrix Composites," in Handbook of Continuous Fiber-Reinforced Ceramic Matrix Composites, R. Lehman et al. (Editors), American Ceramic Society (1995) 355-429.

[13] M. P. Borom, W. B. Hillig, R. N. Singh, W. A. Morrison, and L. V. Interrante, "Fiber containing composites," US Patent 5,015,540, May 1991.

[14] P. G. Karandikar, J.R. Singh, and C. A. Andersson, "Low-expansion metal ceramic composite bodies and methods for making same," US Patent 7,169,465 January 2007.

[15] G. Dietrich, T. Haug, A. Kienzle, C. Schwartz, H. Stover, K. Weisskopf, and R. Gadow, "Fiber reinforced composite ceramics and method of producing the same," US Patent 6,261,981, July 2001.

[16] G. Dietrich, G. Gross, T. Haug, and K. Rebstock, "Brake unit consisting of a brake disk and a brake lining," US Patent 6,079,525, June 2000.

[17] P. G. Collins, and P. Avouris, "Nanotubes for electronics," Scientific American, 283 (6) 2000, pg. 62-69.

[18] M. Treacy, T. W. Ebbesen, and T. M. Gibson, "Exceptionally high Young's modulus observed for individual carbon nanotubes," Nature 381 (1996) pg. 680-687.

[19] E. W. Wong, P. E. Sheehan, C. M. Lieber, "Nanobeam mechanics: elasticity, strength and toughness of nanorods and nanotubes," Science 277 (1997) pg. 1971-1975.

[20] R. S. Ruoff and D.C. Lorents, "Mechanical and thermal properties of carbon nanotubes," Carbon 33 (7) 1995 pg. 925-930.

[21] M. Dresselhaus, G. Dresselhaus, P. Elkund, and R. Saito, "Carbon Nanotubes," Physics World January 1998 (http://physicsweb.org/article/world/11/1/9/1)

[22] E.T. Thostenson, P. G. Karandikar, and Tsu-Wei Chou, "Fabrication and characterization of reaction bonded silicon carbide/carbon nanotube composites," J. Physics D: Applied Physics, Vol. 38, Issue 21, (2005) 3962-3965.

INVESTIGATIONS ON THE STABILITY OF PLATINUM NANOSTRUCTURES ON IMPLANTABLE MICROELECTRODES – A FIRST APPROACH

Anup Ramachandran, Wigand Poppendieck, Klaus Peter Koch
Department of Medical Engineering and Neuroprosthetics, Fraunhofer-IBMT, Ensheimer Strasse-48, 66386 St. Ingbert, Germany

Nicole Donia, Sanjay Mathur
Department of Nanocrystalline Materials and Thin Film Systems, Leibniz-Institute for New Materials, Im Stadtwald, Building D2 2, 66123 Saarbruecken, Germany

ABSTRACT
The size of implantable electrodes for neuroprosthetic application is reduced to the order of micrometers attaining a large number of interconnections. Therefore high surface area materials are employed to reduce the electrode impedance, which helps to achieve good signal-to-noise ratio for the recorded signals as well as to inject high current during stimulation of nerves. Platinum black or platinized Platinum is a material that has been investigated for high surface area applications. The Platinum black can be electrochemically deposited on platinum electrode surfaces from a solution of hexachloroplatinic acid at room temperature. However, the stability of such coatings, especially over a long duration, is not known. This paper presents a first investigation on the stability of Platinum black coating for microelectrode application. The approaches explained here can be used as guidelines for similar investigations on nanostructured coatings on electrodes. The deposition was done over Platinum coated surfaces to study the macroscopic influences of deposition, and over microelectrode structures developed for implantation inside animal model for the investigations. For improving the stability, the platinization was also done in the presence of ultrasound. A comparative study on the ultrasonic and non-ultrasonic Platinum black deposition is presented.

INTRODUCTION

Platinum black (Pt-black) is a conductive material that has been widely used in electrode applications because of its high surface area as a result of its rough surface presentation[1]. Also due to the same reason this material has been investigated for the fabrication of micro-sized and implantable electrodes for neuromuscular stimulation and recording. Generally, Pt-black is deposited over conductive electrode materials such as Platinum, iridium, gold etc., for increasing the surface area in a three-dimensional fashion. The highly chondritic fractal surface of the Pt-black reduces the equivalent current density at the electrode interface, or in other words, there are more sites available for the charge injection.

The deposited Pt-black however tends to flake-off from the metal substrate surface during the application of stress. Various methods have been tried by different groups of investigators to improve the adhesion of Pt-black coating to the substrate. One approach, for instance, uses reversal of polarity during the electrodeposition, so that loosely adhered Pt-black particles will get desorbed from the metal surface whereas strongly adherent particles remain on the substrate to the end of the procedure[2]. In a different strategy deposition was carried out in the presence of ultrasonic agitation in order to achieve well-adhered coatings with enhanced life time[3]. We have deposited Pt-black with and without the presence of ultrasound and also in micro- and macro-sized electrodes. The investigations on microelectrodes revealed that the ultrasound-assisted platinization can be more durable after mechanically stressing the electrodes. Also the stability of the ultrasonically platinized Pt-black when used over a long period of time is not known. Therefore, we also present the impedances of Pt-black coatings on large surface area electrodes for ten days. The deposition of Pt-black on these electrodes was similar to the smaller electrode. The first investigation on the macro-electrodes, however, shows that the ultrasound-assisted coating was comparatively stable over long-term.

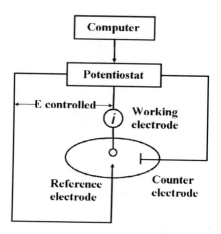

Figure 1: A three-electrode set-up for impedance measurement. The same set up was used for the deposition of Pt-black. Reference electrode: Ag/AgCl; Counter electrode: Platinum with relatively large surface area than the working electrode.

MATERIALS AND METHODS
Pt-black deposition
    The Pt-black coatings were prepared by electroplating under potentiostatic conditions. A mixture of 5 g $H_2PtCl_6*6H_2O$ (HexachloroPlatinum-(IV)-acid Hexahydrate) and 71.4 mg $Pb(NO_3)_2$ (Lead-(II)-Nitrate) in 360 ml distilled water was used as the electrolyte. The electrolytic cell consists of three electrodes and the samples were used as the working electrode (figure 1). During deposition, a voltage (–0.25 V) was applied for 20 s with respect to an Ag/AgCl reference electrode. A Platinum electrode with a large surface area was used as counter electrode.

Stability of ultrasonic Pt-black on microelectrodes
    A polyimide-insulated sieve design that harbors 27 ring electrodes was used to assess the stability of Pt-black coating on microelectrodes. Each electrode on this sieve has an area of 2200 μm^2, fabricated though sputter deposition of Platinum layers (300 nm). Electrical

connections to the electrodes were accomplished by integrated ribbon cables and pads of sputtered gold (300 nm). Design, fabrication and interconnection of these electrodes were discussed elsewhere in detail[4].

Pt-black was electrochemically deposited on two such sieve electrodes with and without ultrasound assistance. For ultrasonic platinization, a sonotrode (model UP100H, Dr. Heilscher) was placed inside the solution, near the electrode and operated at the time of coating (frequency: 30 kHz, power: 50 W). During the coating, the ring electrodes were connected in parallel to provide the same duration of Pt-black coating. The Pt-black coating on the microelectrodes was confirmed by observing the change in color under a light microscope. The electrochemical impedance of each electrode was measured in the frequency range of 10 Hz to 100 kHz before and after Platinum black deposition. The measurements were done using a three-electrode cell that consists of a Platinum counter electrode and an Ag/AgCl reference electrode (Figure 1). The interconnection resistance was small compared to the electrode impedance, thus its influence was negligible.

After Pt-black deposition, the ultrasonically and non-ultrasonically platinized electrodes were both mechanically stressed inside an ultrasonic bath (Laborette-17, Fritsch GmbH) for 15 min in isopropyl alcohol at room temperature (frequency: 35 kHz, power: 80 W) followed by a rinse with deionized water. This process was repeated again for 15 min after the impedances subsequent to the first treatment had been recorded.

Pt-Black long-term assessment

Platinum was sputter-deposited on a glass substrate (layer thickness 300 nm), with a 10 nm titanium layer as adhesion promoter. After contacting the metallization through hand-soldered wires, Platinum electrodes with a surface area of 1×1 cm^2 were prepared from the substrate by encapsulating the remaining areas with a two-component epoxy (UHU plus endfest 300, UHU GmbH&Co KG, Buehl, Germany).

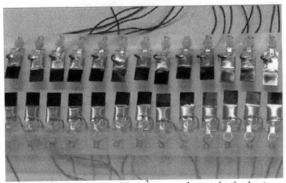

Figure 2: Electrodes with a surface area of 1 cm^2 mounted on a plastic sheet

The electrodes were divided into three groups of 14 each. For the first group, ultrasound was used to improve the adhesion of the Pt-black coating. For the second group, no ultrasound was applied during the coating. The third group remained untreated (bright Pt electrodes).

For long-term electrochemical characterization, the electrodes were first fixed on a plastic sheet (Figure 2), which was then circularly assembled and introduced into physiological saline (0.9 % NaCl at 37°C). The counter electrode consisted of a coiled Platinum wire (0.15 mm diameter), and was positioned in the center of the circle, such that the distance to the counter electrode was approximately equal for all working electrodes. An Ag/AgCl reference electrode was also placed inside the saline solution, to enable a three-electrode configuration.

Impedance spectroscopy was performed during regular intervals over a period of 10 days. The measurement voltage was 50 mV, and the frequency was varied between 0.1 Hz and 100 kHz.

RESULTS AND DISCUSSION

Figure 3 (a) shows a Platinum electrode embedded in the polyimide insulator exposing only the metallization on one side. The mean impedance magnitude of the electrodes was ~163 kΩ at 1 kHz frequency. After the Pt-black deposition (Figure 3 (b)) the mean electrode impedance was reduced to ~ 7 kΩ at 1 kHz frequency because of the increased electrochemically active electrode surface. Figure (c) & (d) shows the morphology of Pt-black over an electrode.

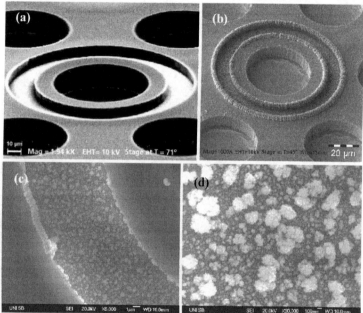

Figure 3: SEM: (a) Platinum ring electrode; (b) After Platinum black deposition with ultrasound; (c) & (d) Magnified view of the Pt-black coating on ring electrode.

Figure 4 shows the impedance magnitude of the electrodes coated with Pt-black through electrochemical means, without the use of ultrasound.

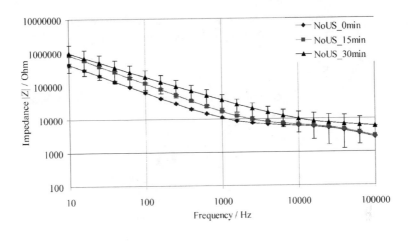

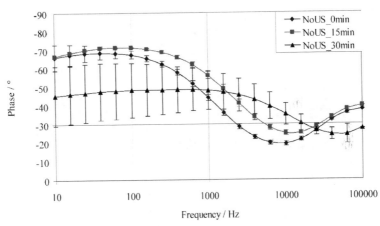

Figure 4: Impedance magnitude (above) and phase (below) of the non-ultrasound-assisted Pt-black electrodes and increase in impedance after mechanical stress inside isopropyl alcohol for 15 min and 30 min.

For example, if we observe the mean impedance magnitude of these electrodes at 1 kHz frequency the impedance was approximately 10.7±0.8 kΩ (mean value±standard deviation) after deposition. After the application of ultrasonic stress for 15 min the impedance was increased to 16.7±1.4 kΩ. The measurement after 30 min showed a further increase in the impedance values to 36.6±23 kΩ. The standard deviation of the electrodes after 30 min of ultrasonic stress was considerably high. The destruction of coating due to the flaking off of the Pt-black layers from some of the electrode sites was observable under light microscope.

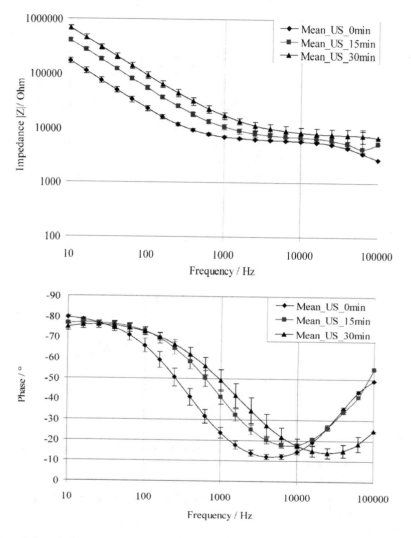

Figure 5: Impedance magnitude (above) and phase (below) of the ultrasound-assisted Pt-black electrodes and increase in impedance after mechanical stress inside isopropyl alcohol for 15 min and 30 min.

On the other hand, the impedance magnitude of the ultrasonically platinized electrodes was initially 6.9 ± 0.3 kΩ (Figure 5). After the 15 min and 30 min stress an increase in the impedance occurred, up to values of 10.5 ± 1.1 kΩ and 17 ± 3.5 kΩ. When we compare both batches of electrodes, the ultrasound-assisted Pt-black electrodes are found to be more stable against mechanical stress. The comparison of the first measurements of these electrodes also showed a lower impedance magnitude and less standard deviation compared to the non-ultrasonically platinized electrode. The reason for this could be the elimination of the surface impurities because of the ultrasound agitation during the deposition.

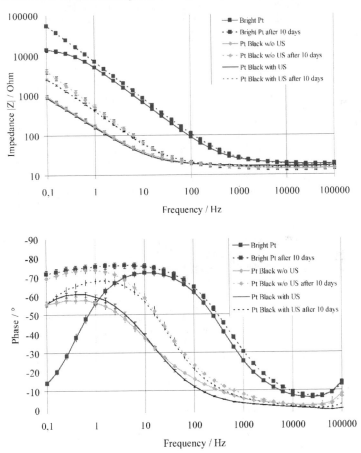

Figure 6: Results of the long-term impedance spectroscopy (above: absolute impedance, below: phase). Average values for the three groups (bright Pt, Pt-black w/o ultrasound, Pt-black with ultrasound) at the beginning of the measurements and after ten days are presented.

In Figure 6, the results of the long-term impedance spectroscopy with the 1x1 cm² electrodes are given. Average values for the three groups (bright Pt, Pt-black deposited without the use of ultrasound, and Pt-black deposited with ultrasound) at the beginning of the measurements and after ten days are presented with respective standard deviation (n=14). The impedance of all the three groups of electrodes was increased after 10 days inside 0.9 % NaCl solution. This could be also because of the reduction in area due to self diffusion of Platinum and Pt-black surfaces[2]. However, the impedance of Pt-black coated electrodes was less than the bright electrodes especially in the low frequency range. Further, it can be seen that the ultrasonically prepared Pt-black electrodes exhibit lower impedance in the low-frequency range after ten days, compared to the Pt-black electrodes prepared without using ultrasound. The trend indicates that the ultrasonic treatment provides a better long-term stability for Pt-black deposition.

SUMMARY AND CONCLUSIONS

Quantitative data and methods of investigation addressing the stability of surface coatings such as Pt-black have not been sufficiently available, especially in the case of microfabricated electrodes. Nevertheless, Pt-black deposition was found to be useful in case of small electrode dimensions where the impedance is limited by the geometrical constraints of the design. In this case the Pt-black grows rather in a three-dimensional manner, increasing the real surface area. Also the methods adopted here could be extended and can be used for similar investigations on nanostructure coatings. However, the choice of using the Pt-black coating for implantable microelectrode application also depends on the biocompatibility of this material[5] and further investigations are necessary to qualify Pt-black for implantable long-term applications. Finally it is a compromise for the high surface area and the long-term stability of such coatings.

ACKNOWLEDGEMENTS

This research is supported by the European Union projects: NEUROBOTICS (IST-FET-2003-001917) and ROSANA (IST -2001-34892). The authors express their sincere thanks to Dr. Hao Shen, Dipl.-Ing. Thomas Doerge and Evgueni Kounik for the SEM pictures.

REFERENCES
[1]Ilic, B., Czaplewski, D., Neuzil, P., Stanczyk, T., Blough, J., Maclay, J.G., "Preparation and Characterization of Platinum Black Electrodes", Journal of Mat. Sci., 35, 3447-57 (2000)
[2]Feltham, A. M., Spiro, M., "Platinized Platinum Electrode", Chemical Reviews, 71, pp. 117-93 (1970)
[3]Marrese, C.A., "Preparation of Strongly Adherent Platinum Black Coating", Anal. Chem., 59, 217-18 (1986)
[4]Ramachandran, A., Schuettler, M., Lago, N., Doerge, T., Koch, K. P., Navarro, X., Hoffmann, K. P., Stieglitz, T., "Design, in vitro and in vivo assessment of a multi-channel sieve electrode with integrated multiplexer", Journal of Neural Engineering, 3, 114–24 (2006)
[5]Schuettler, M., Doerge, T., Wien, S,L., Becker, S., Staiger, A., Hanauer, M., Kammer, S., Stieglitz, T., "Cytotoxicity of Platinum Black", Proc. 10th Annual IFESS conference, Canada, july (2005)

GROWTH OF BARIUM HEXAFERRITE NANOPARTICLE COATINGS BY LASER-ASSISTED SPRAY PYROLYSIS

G. Dedigamuwa, P. Mukherjee, H. Srikanth, and S. Witanachchi*
Department of Physics, University of South Florida, Tampa, 33620

ABSTRACT

Barium Hexaferrite (BaM) nanoparticle coatings have been deposited by a laser assisted spray pyrolysis method. The concentration of the Ba and Fe inorganic salts in the starting aqueous solution determined the final particle size. Films with particle size tunability in the range of 30-100 nm have been grown. Incorporation of CO_2 laser heating in the process enabled the growth of well-defined smaller particles in comparison to thermal pyrolysis. The use of SF_6 as a carrier gas, which has a high absorption at the CO_2 laser wavelength, heated the nebulized droplets above $150^\circ C$. Several post annealing conditions to form the $BaFe_{12}O_7$ structure have been studied. Structural and morphological characterization of the films has been carried out by x-ray diffraction, Scanning Electron Microscopy (SEM), and Atomic Force Microscopy (AFM). Magnetic properties of the nanoparticle films have been investigated by using a Physical Property Measurement System (PPMS).

INTRODUCTION

The formation of films consisting of particles of magnetic materials with size tunability is important for applications that require single magnetic domains. Some of the applications include magneto-sensors, bio-sensors, magneto-electronics, data storage, magnetic heads of computer hard disks, single-electron devices, microwave electronic devices, etc[1,2]. Out of the variety of magnetic materials available, barium hexaferrite[3] is one of the most widely studied materials due to its unique properties, that include high uniaxial magnetic anisotropy, high Curie temperature, high coercive force, and large saturation magnetization. Most of the applications require the grains to be small enough to contain only a single magnetic domain while large enough not to be superparamagnetic. As a result, several different techniques have been developed to grow nanoparticles and nanograined films of barium hexaferrite[1].

One of the simplest methods for the preparation of BaM is firing mixtures of iron oxide and barium carbonate at high temperatures ($1200^\circ C$) followed by grinding to reduce the particle size[4]. This method generally yields mixtures which are non-homogeneous. In addition, BaM nanoparticle powders have also been grown by several methods that include chemical coprecipitation[5,6], sol-gel method[7], citrate precursor method[8], and thermal spray pyrolysis[9]. The citrate precursor method has the advantage of forming the BaM phase at a post annealing temperature of about 700C in comparison to $900^\circ C$ required by other techniques[10]. In this method aqueous solutions of stoichiometric amounts of Ba^{2+} and Fe^{3+} nitrates are reacted with citric acid in a controlled pH environment to form a citrate complex. Thermal decomposition of the complex around $700^\circ C$ leads to the formation of BaM.

The citrate precursor method has also been combined with spray pyrolysis to form nanoparticle coatings of BaF^{11}. The chemical spray pyrolysis technique can be used to

*Corresponding author, e-mail: switanac@cas.usf.edu

form coatings of a variety of different materials[13]. The main component of a spray pyrolysis system is an atomizer that generates micro-droplets of a precursor solution dissolved in a relatively volatile solvent. Ultrasonic nebulizers are known to produce a fairly uniform distribution of micrometer size droplets. Generally, the nebulizer is operated at a frequency of 2.4 MHz, where the precursor solution is converted into a mist of particles in a range of 1-2 μm in diameter. These particles lack sufficient inertia and thus have to be transported by a carrier gas. The droplets in the form of a fine spray are carried out of a nozzle onto a heated substrate by the carrier gas that can be inert or reactive. The constituents of the droplets decompose and react on the hot substrate to form the chemical compound. The substrate temperature should be high enough to evaporate the volatile solvents.

The precursor is decomposed to form the compound when the droplets impinge upon the heated substrate. The subsequent film formation and morphology of the film is dependant on the velocity of the drop, rate of reaction and the rate of evaporation of the solvent. At high velocities the droplets will flatten on the substrate leading to large particle sizes[11]. If the solvent in a droplet is evaporated so that the salts are condensed into a particle, the particle tends to retain its shape when it strikes the substrate. We have developed a laser-assisted spray pyrolysis method in which the evaporation of the solvent by the laser leads to the deposition of nanoparticle films with particle sizes that are smaller than those deposited without the laser. The solvent in the droplets are evaporated at the nozzle by a CO_2 laser that causes the material in the droplets to condense into smaller solid particles. The laser energy is transferred to the droplets by the carrier gas SF_6 which has a high absorption at the CO_2 laser wavelength. The size of the particles deposited on the substrate can be controlled by controlling the concentration of the solvent. The role of the laser in the laser assisted spray pyrolysis process can be seen in the AFM images of films deposited with and without the laser (Fig. 1).

Formation of the BaM phase and the particle size distribution of the nanoparticle coatings have been observed to depend on the composition and the concentration of the starting solution as well as the post annealing condition. Several reports have pointed out the

(a)   (b)

Fig. 1: Films deposited by spray pyrolysis method (a) without laser heating, and (b) with laser heating

need for excess Ba to form the BaM phase[13]. Even though the citrate precursor method facilitates the formation of BaM at a lower annealing temperature, time of annealing should be carefully controlled to prevent grain growth. This paper presents the results of a systematic study that was carried out to determine the optimum growth parameters to form nanoparticle BaM coatings with particle size control.

EXPERIMENTAL METHOD

Aerosols with an average drop size of 1.5 µm were generated by a (Sonaer Ultrasonic Model 241CST) nebulizer. The aerosol was carried through a conical nozzle into the growth chamber by SF_6 gas. The schematic diagram of the laser-assisted spray pyrolysis system is shown in Fig. 2. The droplets and the SF_6 gas interact with the focused 10.6 µm wavelength CO_2 laser beam at the nozzle. The laser power of 3W used in the experiment was sufficient to increase the temperature of the gas and subsequently the droplets to about $350^{\circ}C$. Approximate value of the gas temperature was measured in a calibration experiment where a thermocouple was placed in front of the gas jet, just outside the laser beam-gas interaction volume. Oxygen was introduced into the growth chamber to promote oxidation of the depositing film. The flow rate and thus the speed of the aerosol into the chamber were controlled by gas flow meters. The pressure inside the chamber was about 700 Torr. Films were deposited on quartz or silicon substrates at a temperature of $300^{\circ}C$. The growth rate of the film depended on the concentration of the precursor. The average growth rate was about 50 nm/min.

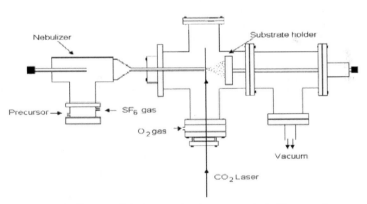

Fig. 2: Schematic diagram of the laser-assisted spray pyrolysis film growth system.

Two different precursors were studied using the laser-assisted spray pyrolysis method. The first precursor was prepared by dissolving BaI_2 and $FeCl_3$ in water. In addition to the stoichiometric Ba:Fe ratio of 1:12, films were deposited with Ba:Fe ratios of 1:6, 1:4, and 1:3. All the films were post annealed in air in a furnace using three different schemes: (a) anneal at

1000 °C for 5 minutes, (b) increase the temperature to 1000°C at the rate of 10°C/min followed by 10°C/min decrease to room temperature with no holding time, (c) increase the temperature to 1000°C at the rate of 10°C/min, hold for 1 hour, followed by 10°C/min decrease to room temperature.

The second precursor was prepared by the citate method. The aqueous solution of 0.25M concentration Ba^{2+} and Fe^{3+} citrate precursor was made by dissolving $Fe(NO_3)_3.9H_2O$ (99% purity, Alfa Asaer), $Ba(OH)_2.8H_2O$ (98% purity, Alfa Aesar), and anhydrous citric acid (99.5% purity, Alfa Aesar) in molar ratios of $Ba^{2+}:Fe^{3+}:C_3H4(OH)(COOH)_3=1:12:19$ in de-ionized water. The solution was then neutralized (pH=7) using an aqueous solution of NH_3 (concentration: 28 wt%). The neutralized solution was refluxed at 60 °C for 3 h to allow carboxyl groups of citric acid to completely chelate metallic ions in the solution. Films deposited by using this precursor were post annealed at a temperature of 700°C with a heating and cooling rate of 10°C/min and a holding time of 1 hour. Films were also deposited using solutions with concentrations of 0.2M and 0.1M.

The stoichiometry of as-deposited films was determined by EDS analysis. Crystallinity of the films was analyzed by x-ray diffraction while the morphology of the films and the nanoparticle sizes were determined by SEM and AFM. The magnetic properties of the BaM film grown under optimum conditions were investigated by a PPMS with a maximum magnetic field of 7 T.

RESULTS AND DISCUSSION

EDS analysis of the as deposited films from aqueous solutions of BaI_2 and $FeCl_3$ (first precursor) for different Ba:Fe ratios agreed closely with the composition of the starting solution. X-ray diffraction results of films deposited with this solution for the Ba:Fe ratios of 1:12, 1:6, and 1:3 are shown in Fig. 2. All the films have been post annealed in oxygen at 1000°C for 2 hours. X-ray diffraction of the films deposited with stoichometric composition of the starting solution (1:12) showed very weak signals from the BaM phase while preferential formation of Fe_2O_3 is indicated by the presence of strong peaks corresponding to Fe_2O_3. With increasing Ba content suppression of the Fe_2O_3 phase and formation of BaM was observed. For a composition ratio of 1:3, which contained 300% more Ba than what is present in a stoichiometric solution, only the BaM phase was observed. This result is consistent with other reports that point out the requirement of excess Ba to stabilize the BaM phase[14]. In a typical x-ray powder pattern of BaM the highest intensity peaks are observed for the (107) and (114) orientations. Intensity ratio between the (008) and (107) is about 1:5. However, the highest peak observed for the deposited films was the (008) orientation, which indicates that most of the particles in the film are deposited on the substrate with a c-axis orientation.

AFM image of the film (Fig. 4) clearly show the hexagonal grains which are c-axis oriented. It is also clear from this image that grain growth during the annealing process have given rise to grain sizes in the micron range. To prevent grain growth annealing temperature and annealing time have to be lowered. An SEM image of a film annealed at 1000°C for 5 minutes is shown in Fig. 5. The grain sizes are of the order of 200 nm. However the structure appears to be cubic. Formation of the BaM structure under different annealing conditions was studied by x-ray analysis of the films (Fig. 6). According to Fig. 6 (b), the 5 minute annealed film show only the Fe_2O_3 structure, which is consistent with the cubic structure observed in Fig. 5. These results suggest that gradual increase of temperature to 1000°C is important for the formation of the BaM Phase. When the films are held at 1000°C for a longer time, larger grains are formed producing high x-ray intensities. Films deposited by the citrate precursor were also analyzed by x-ray diffraction. Fig. 7 shows diffraction patterns corresponding to an as deposited film and a film annealed at 700°C for 1 hour.

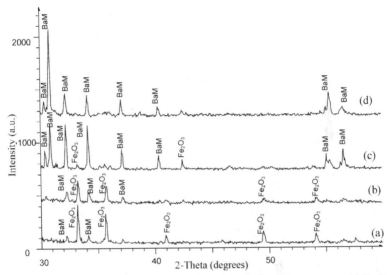

Fig. 3. X-ray diffraction patterns of films deposited using precursors with Ba:Fe ratios of (a) 1:12, (b) 1:6, (c) 1:4, and (d) 1:3. Films were post annealed at 1000C for 1 hour.

Figure 4. AFM image of film (d), which has only BaM peaks.

Figure 5. SEM image of a film annealed for 5min. at 1000 °C

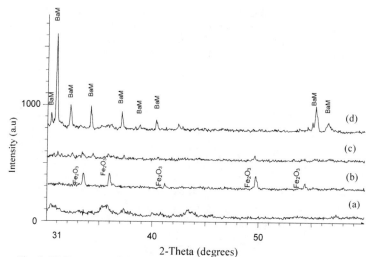

Fig. 6. XRD patterns of the samples that are subjected to different heat treatments. (a) As deposited. (b) Quick annealed (5 mins) at 1000 °C. (c) Ramp up to 1000 °C and ramp down. (d) Ramp up to 1000 °C, soak 1 hrs at 1000 °C and ramp down.

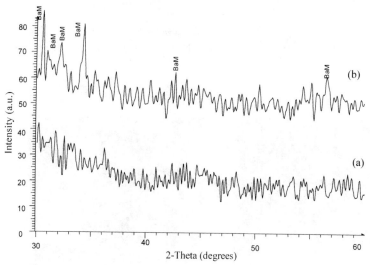

Fig. 7: X-ray diffraction patterns of films deposited with the citrate precursor (a) as deposited, (b) annealed for 700 °C.

All the peaks observed in the annealed film correspond to the polycrystalline structure of BaM. As indicated by the AFM images (Fig. 8) of films deposited by the citrate

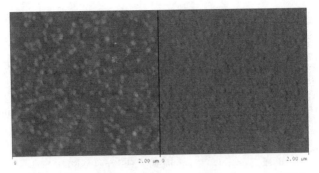

Fig. 8 : AFM image of the film deposited with a precursor of 0.25 M concentration and annealed for 700 °C.

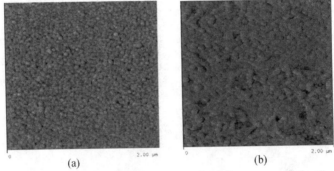

(a)   (b)

Fig. 9: 2D AFM image of the Samples prepared by different concentration at substrate Temp 350 °C by spray pyrolysis with the laser heating. (a) 0.1M, estimated average particle size: 50 nm, (b) 0.2M, estimated average particle size: 65 nm.

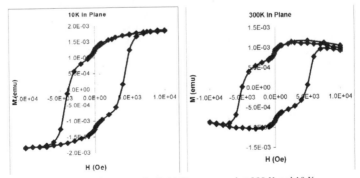

Fig. 10: Hyseteresis curves of a BaM film measured at 300 K and 10 K.

precursor method, the particle size remained in the nanoscale after the annealing step. The average particle size was about 50 nm. Fig. 9 shows AFM images of films deposited by the citrate precursor and post annealed at 700°C for starting solution concentrations of 0.2M and 0.1M. The average particle sizes for these films were 50 nm and 40 nm, respectively. X-ray diffraction of these films also showed diffraction peaks that are similar to those in Fig. 8.

M-H curves measured using a Physical Property Measurement System (PPMS) for films deposited by the citrate precursor and annealed at 700°C for 1 hour are shown in Fig. 10. Based on the measured hysteresis curves for magnetic fields applied perpendicular to the film, high coercivities of more than 4000 Oe were observed in both room temperature and at low temperature. The large coercivity indicates that the nanograins are mostly c-axis oriented.

CONCLUSION

We have compared two different precursors for the growth of BaM nanoparticle film using a laser assisted spray pyrolysis technique. This method has been successful in reproducing the stoichiometry of the starting solution in the deposited films. All the films had to be post annealed to form the BaM phase. When an aqueous solution of a mixture of BaI_2 and $FeCl_3$ was used as the precursor, stoichiometric solutions led to the formation of Fe_2O_3. Excess Ba in the solution, up to a Ba:Fe ratio of 1:3, was required to form the BaM structure. Furthermore, the required high temperature annealing step at 1000°C caused grain growth leading to grains in the micrometer range. The use of stoichiometric citrate precursors in laser assisted spray pyrolysis produced the BaM phase at an annealing temperature of 700°C. Low annealing temperature prevented grain growth leading to nanoparticles whose size was determined only by the concentration of the starting solution. A majority of the nanograins are preferentially oriented along the c-axis. Large coercivities (in excess of 4000 Oe) observed in hysteresis curves also indicate the c-axis orientation of the films.

ACKNOWLEDGEMENTS

This work was partially supported by the National Science Foundation, under Grant No. DMI-0127939.

REFERENCES

[1] B.J. Palla, D.O. Shah, P. Garcia-Casillas and J. Matutes-Aquino, "Preparation of Nanoparticles of Barium Ferrite from Precipitation in Microemulsions"., *Journal of Nanoparticle Research.*, Vol 1, 2, 215-221 (1999).

[2] Kubo, O. Ido, T.Yokoyama, H., "Properties of Ba ferrite particles for perpendicular magnetic recording media.", *Magnetics IEEE*, Vol 18, issue 6, 1122-1124 (1982).

[3] C. S. Kim, S. W. Lee, S. Y. An, In-Bo Shim, "Magnetic Properties of Barium Ferrite Thin Films on Pt(111) by a Sol–Gel Method" *phys. stat. sol.*, (a) 189, 3, 903–906 (2002).

[4] G. A. Ward, K. H. Sandhage., "Synthesis of Barium Hexaferrite by the Oxidation of a Metallic Barium–Iron Precursor"., *J. American Ceramic Society.*, 80, 6, 1508 (1997).

[5] T. Ogasawara, M.A.S. Oliveira,."Microstructure and hysteresis curves of the barium hexaferrite from co-precipitation by organic agent.", *J. Magn. Magn. Mater.*, 217, 72 (2000).

[6] S.R. Janasi, M. Emura, F.J.G. Landgraf, D. Rodrigues,. "The effects of synthesis variables on the magnetic properties of coprecipitated barium ferrite powders.", *J. Magn. Magn. Mater.*, 238, 2-3, 168-172 (2002).

[7]W. Zhong, W. Ding, Y. Jiang, N. Zhang, J. Zhang, Y. Du, Q. Yan., "Preparation and Magnetic Properties of Barium Hexaferrite Nanoparticles Produced by the Citrate Process"., *J. American Ceramic Society.*, **80**, 12, 3258 (1997).

[8]V. K. Sankaranarayanan, R. P. Pant and A. C. Rastogi, "Spray pyrolytic deposition of barium hexaferrite thin "lms for magnetic recording applications"., *J. Magn. Magn. Mater.*, **220**, 72-78 (2000).

[9]M.V. Cabanas, J.M. Gonzalez-Calbet, M. Labeau, P. Mollard, M. Pernet, M. Vallet-Regi., "Evolution of the microstructure and its influence on the magnetic properties of aerosol synthesized BaFe[sub 12]O[sub 19] particles" *J. Solid State Chem.*, **101**, 265 (1992).

[10]E. E. Vidal, P. R. Taylor, "Fundamental consideration in the deposition of Barium ferrite films in an inductively coupled plasma reactor"., *Annals of the New York Academy of Sciences* **891**, 152-163 (1999).

[11]Hsuan-Fu Yu and Hsin-Yi Lin,. "Preparation and thermal behavior of aerosol-derived BaFe12O19 nanoparticles"., *J. Magn. Magn. Mater.* **283**, 190-198 (2004).

[12]K. Okuyama, L. Wuled, N. Tagami, S. Tamaki and N. Tohge,. "Preparation of ZnS and CdS fine particles with different particle sizes by a spray-pyrolysis method"., *J. Mat. Sci.* **32**, 5,1229 (1997).

[13]R. C. Pullar, A. K. Bhattacharya., Crystallisation of hexagonal M ferrites from a stochiometric sol-gel precursor, without formation of the α-BaFe$_2$O$_4$ intermediate phase.

[14]Y. Goto, T. Takada., Phase diagram of the system BaO-Fe2O3"., J. Am. Cream. Soc. 43, 150, (1960).

CRACK EXTENSION BEHAVIOR IN NANO-LAMINAR GLASS/METAL COMPOSITE

Hideki Kakisawa, Taro Sumitomo
Composites & Coatings Center, National Institute for Materials Science
1-2-1, Sengen, Tsukuba, Ibaraki 305-0047 Japan

Yusuke Owaki
Department of Materials Engineering, Graduate School, University of Tokyo
4-6-1, Komaba, Meguro-ku, Tokyo 153-8904 Japan

Yutaka Kagawa
Research Center of Advanced Science and Technology, University of Tokyo
4-6-1, Komaba, Meguro-ku, Tokyo 153-8904 Japan

ABSTRACT
     A bulk nano-laminar ceramic composite was fabricated by a simple sintering technique. Thin glass flake powder coated with either nickel or silver was used as the raw material, and was sintered by hotpressing. The optimum sintering condition for each material was investigated. Samples fabricated in the optimum condition had a dense, laminar microstructure originating from the aligned flake powder. Indentation tests of the samples suggested their high resistance to crack propagation through the transverse direction of the lamellar; this result was attributed to crack deflection at the interface and accumulation of microfractures around the indentation. The result of a three-point bending test for the samples, when the powder was aligned in advance during the green sample fabrication, showed a stable fracture after the maximum load, while the samples fabricated by simple hotpressing of the powder without pre-alignment fractured unstably.

INTRODUCTION

     Laminar composites have been attracting attention because of their apparently high fracture resistance[1-10]. The damage tolerance of laminar composites is achieved through crack arrest by the laminates and deflection at the laminates' interfaces. This results in a complicated fracture behavior with a large failure strain before the final fracture, and avoids the catastrophic fracture common to monolithic ceramics. Recently, researchers inspired by natural composites like bones and shell have attempted to develop laminar composites with ultra-thin laminates, at less than the micron order[11-15].
     Advanced thin film processing methods have produced nanometer-thin layers: for example, self-assembly by biomimetic processing[16], the layer-by-layer method,[12, 13] chemical and physical vapor deposition (CVD and PVD)[15, 17], spin coating[8],19], etc. These have the potential to form very thin, uniform films with a controlled thickness, and are expected to produce coating materials and functional materials for electronic and magnetic applications. However, these methods are not suited for fabrication of bulk material for structural use. A simple and reasonable processing method is in demand.
     In this paper, a process using a powder metallurgical technique is used for fabricating such a nano-laminar ceramics composite. Specifically, thin flake powder coated with an interface

material is aligned and sintered. Model composites were fabricated from glass flake powder with metal coating.

EXPERIMENTAL PROCEDURE

Either alkali glass powder with nickel coating (Metashine MC05040N: Nippon Sheet Glass Co., Ltd., Tokyo) or non-alkali glass powder with silver coating (Metashine MEG040S) was used as the raw material. The powder had a flake-like shape with a nominal size of 40μm, as shown in Fig. 1. The thickness of the nickel-coated powder and the silver-coated powder was 2μm and 0.7μm, respectively. The powder was coated by electroless plating, and the thickness of the coating was about 50nm. The properties of the glass, nickel and silver are summarized in Table 1. [20], [21]

20μm

Fig. 1 Appearance of the glass flake.

The powder was put into a graphite die of 40×30mm and hotpressed for 3.6ks with a pressure of 40MPa in a vacuum. Hotpressing temperatures of 873, 923, 953, 973, 993, 1023 and 1073K were tested. Hereafter, the samples from the silver-coated glass are designated S-(hotpressing temperature), e.g., S-1023. The samples made from the nickel-coated powder are designated N-(hotpressing temperature), e.g., N-923.

Some of the samples were fabricated with a more complicated processing using a 3D printing machine (Z402, Z Corporation, USA). Figure 2 shows a schematic illustration of the 3D printing process. A 3D CAD model of specimen was converted to STL format and then sliced into a stack of thin layers. The data of each layer was sent to the 3D printing machine. Powder for a layer was carried from the powder-supplying stage and spread over the building stage by a

Table 1 Properties of the glass and metals[20], [21]

	Alkali glass	Non-alkali glass	Nickel	Silver
Young's modulus (GPa)	73	69	207	71
Density (g/cm^3)	~2.6	~2.5	8.9	10.5
Glass softening temperature (K)	1113	1023		-
Melting Temperature (K)	-	-	1726	1235

roller. The cross-section of the layer, in this case a rectangle of 40×30mm, was printed with polyvinylalcohol (PVA) binder in the building stage according to the data sent (see (1) in Fig. 2). Then the building stage was lowered by the layer thickness, and new powder for the next layer was spread over the previous layer ((2) and (3) in Fig. 2). The serious of the process was repeated and a green sample of 40×30×~20mm was fabricated. The layer thickness was set at 90μm. Two green samples were stacked in the graphite die and hot pressed at 1023K. Hereafter, this sample is designated S-1023-3D.

The sintered samples were cut parallel and perpendicular to the laminating direction with a diamond saw and polished using a diamond 0.05μm alumina suspension for the microstructure observation by optical and scanning electron microscopy (SEM). A Vickers indentation test of N-923 and S-1023 was carried out on the polished surfaces using a standard pyramidal indenter for 40s at an indentation load of 49-196N. The indentation was oriented carefully so that one of the diagonal lines of the permanent impression would be parallel with the lamination direction. After indentation, the crack propagation behavior near the indentation mark was examined.

A three-point binding test of S-1023 and S-1023-3D was done with a crosshead speed of 0.05mm/min. The dimension of the specimens was 3×15×4 mm (width×span×depth). The specimen preparation was done according to the Japanese industrial standard (JIS).

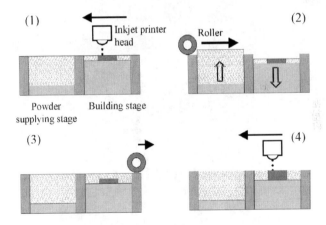

Figure 2 Schematic illustration of 3D printing procedure.

RESULTS AND DISCUSSION

The density of the samples increased with increases in the hot pressing temperature. High temperature allowed the deformation of glass powders and adhesion of the coating metal. In this study, full density (over 99% of the theoretical) was achieved in S-1023 and N-923. At higher temperatures, cohesion of the coating metal occurred, into the spherical inclusions. Thus, the optimum hotpressing temperature was determined to be 1023K for the silver-coated powder and

923K for the nickel-coated powder. The difference in the optimum temperature was mainly due to the difference of the glass-softening temperature: a temperature about a hundred degrees below the glass-softening temperature was required for full densification in this study.

Figure 3 shows the cross-sections of the samples hot-pressed at the optimum conditions: (a) N-923, (b) S-1023, and (c) S-1023-3D. The dark parts are the glass phase and the bright parts are the metal phase. The individual powder flexibly bent and filled the spaces; few pores were observed in all the samples. The powder was aligned uniaxially: The longitudinal direction of the powder was perpendicular to the loading direction, indicating that the lamellar alignment was achieved by uniaxial press. The coating remained on the powder surface, forming interface layers between the glass plates. In the sample from the nickel-coated powder, N-923 (Fig. 3(a)), the coating layer became less than 10nm, and some parts were discontinuous. Some chemical reaction of glass and nickel or diffusion of nickel to the glass phase might occur though no reaction phase was observed in SEM observation. The samples from the silver-coated powder (Fig. 3(b) and (c)) had an interface layer about 100-200 nm thick. The sample from the green sample by 3D printing, S-1023-3D, had better alignment, while the glass powder bonded with itself in places in the sample that was simply hot pressed from raw powder, or S-1023.

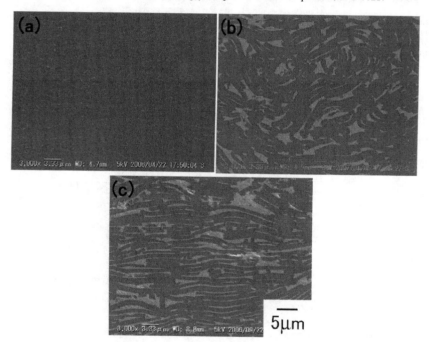

Fig. 3 SEM observation of the cross-section of the samples: (a) from nickel coated powder (N-923), (b) from silver coated powder (S-1023) and (c) from the green sample fabricated from silver coated powder by 3D printing (S-1023-3D).

Figure 4 shows the Vickers indentation of N-923 and S-1023. In N-923, cracks propagated from the corners of the indention both across and along the lamellar directions (Fig. 4(a)). In the direction parallel to the lamellar, more than one crack was observed, and they were combined with each other under the specimen surface. Detailed observation of the crack extension showed that the crack continued straight without deflection, as shown in Fig. 4(c). The thin nickel layers suggested that strong bonding occurred at the interface by some kind of reaction or diffusion, prohibiting the debonding of the interface, crack deflection, or crack arrest at the interface layer.

In S-1023, no crack passing through the lamellar was observed in optical microscopy at any load up to 196N (Fig. 4(b)). Along the lamellar direction, cracks propagated from the sides of the rhombus as well as from the corners. Crack length and crack opening displacement increased with the increase of the load. Observation near the indentation by SEM indicated accumulated microfractures around the indentation. The glass plates were broken at the edges of the indentation (shown by dotted lines in the figure), and cracks were propagated from the edges in the lamellar direction, deflecting along the interface layers. Near the corner oriented in the transverse direction, short transverse cracks of sub-micron size to several micrometers were accumulated within the area 10-20μm from the corner as shown in Fig. 4(d). Most of the cracks

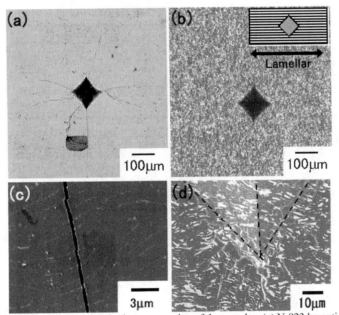

Fig. 4 Vickers indentation on the cross section of the samples: (a) N-923 by optical microscopy, (b) S-1023 by optical microscopy, (c) Crack extension behavior across lamellar in N-923, and (d) accumulated microfracture around indentation corner in S-1023.

passed through one or two glass laminates and stopped at the silver layer; some were deflected and propagated along the interface. Thus, the accumulation of only very small fractures occurred.

Figure 5 shows the relation between load and displacement in the three-point bending test of the samples from the silver-coated powder, S-1023 and S-1023-3D. S-1023 showed a linear load-displacement curve up to the maximum load and then fractured unstably. The crack path was almost flat with some facets, suggesting that the fracture was nearly brittle. SEM observation of the fracture surfaces showed that the crack propagated preferably at the glass-rich regions, the regions in which the glass directly bonded with itself. In S-1023-3D, the load increased linearly, followed by a gradual decrease after the maximum load. The crack propagated in a slanting direction at the maximum load and then proceeded stably in a zigzag fashion. The fracture surface in SEM observation was quite complicated; the crack passed through the interface between the glass, repeating the deflection at the crack tip and the following pullout of the glass plate behind the crack tip.

From those results, the composite showed possibilities for good resistance against crack

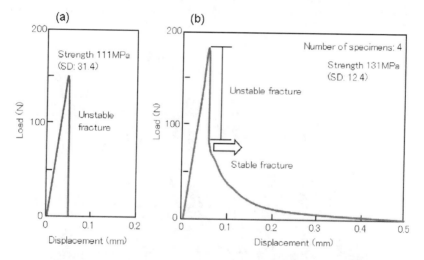

Fig. 5 Load- displacement curve in three point bending tests of the samples from the silver coated powder: (a) the sample fabricated from the green sample and (b) the sample from the powder.

propagation in the transverse direction. The fracture behavior was influenced by the microstructure on the submicron order; the design of the powder alignment and the determination of the processing condition to retain the glass/metal layered structure were the keys to obtaining fracture resistance. This study suggests that the simple method of coating flakes and sintering them in an aligned way can be a good solution to fabricating a nano-laminar composite with damage-tolerant properties.

CONCLUSION

1) A nano-laminar composite was successfully fabricated by hotpressing glass powder coated with metal. The composite had a fully dense, roughly laminated microstructure, and the powder was aligned perpendicularly to the pressing direction. The coating remained at the powder surface, forming an interface layer between the glass plates.
2) In the Vickers indentation test, cracks from the corner of the indentation passed across the interface, probably because of the strong interface in the samples fabricated from nickel-coated powder. No crack propagated in the transverse direction in the samples from the silver-coated powder.
3) The three-point bending test of the samples from the silver-coated powder showed the importance of the powder alignment. The sample from the green sample, which had better alignment, showed a stable fracture with a complicated fracture path.
4) The possibilities of fabricating a nano-laminar composite by a simple powder metallurgical method have been shown.

ACKNOWLEDGEMENT

This work was supported by a Grant-in-Aid for Scientific Research from the Ministry of Education, Culture, Sports, Science and Technology of Japan (No.16760555).

REFERENCES

1) H.M.Chan, "Layered Ceramics: Processing and Mechanical Behavior," Annu. Rev. Mater. Sci. 27, 249-282 (1997).
2) M.P.Harmer, H.M. Chan and G.A. Miller, "Unique Opportunities for Microstructural Engineering with Duplex and Laminar Ceramic Composites," J. Am. Ceram. Soc. 75 (7), 1715-1728 (1992).
3) W.J. CLEGG, "The Fabrication and Failure of Laminar Ceramic Composites," Acta Metall. Mater. 40, 3085-3093 (1992).
4) M. Oechsner, C. Hillman, and F.F. Lange, "Crack Bifurcation in Laminar Ceramic Composites," J. Am, Ceram. Soc. 79, 1834-1838 (1996).
5) T. Ohji, Y. Shigegaki, T. Miyajima, and S. Kanzaki, "Fracture Resistance Behavior of Multilayered Silicon Nitride," J. Am, Ceram. Soc. 80, 991-994 (1997).
6) B. Hatton, and P.S. Nicholson, "Design and Fracture of Layered $Al_2O_3/TZ3Y$ Composites Produced by Electrophoretic Deposition," J. Am, Ceram. Soc. 84, 571-576 (2001).
7) J.L.Huang, Y.L. Chang, and H.H. Lu, "Fabrication of Multilaminated Si_3N_4/TiN Composites and Its Asisotropic Fracture Behavior," J. Mater. Res. 12, 2337-2344 (1997).
8) W.J.Phillips, W.J. CLEGG, and T.W. Clyne, "Fracture Behavior of Ceramic Laminates in Bending-I. Modeling of Crack Propagation," Acta Metall. Mater. 41, 805-817 (1993).
9) C. You, D.L. Jiang and S.H. Tan, "Deposition of Silicon Carbide/Titanium Carbide Laminar Ceramics by Electrophoresis and Densification by Spark Plasma Sintering," J. Am, Ceram. Soc. 87, 759-761 (2004).
10) M.G. Pontin, and F.F.Lange, "Crack Bifurcation at the Surface of Laminar Ceramics That Exhibit a Threshold Strength" J. Am, Ceram. Soc. 88, 1315-1317 (2005).

11) H. Kakisawa, K. Minagawa, Y. Osawa, and S. Takamori, "Fabrication of Nano-Laminar Composite from Glass Flake," J. Ceram. Soc. Jpn. 113, 808-811 (2005).

12) Y.Lvov, F. Essler and G. Decher, "Combination of Polycation/Polyanion Self-Assembly and Langmuir-Blodgett Transfer for the Construction of Superlattice Films," J. Phys. Chem. 97, 13773-13777 (1993).

13) P. Podsiadlo, S. Paternel, J.M. Rouillard, Z. Zhang, J. Lee, J.W. Lee, E. Gulari and N.A. Kotov, "Layer-by-Layer Assembly of Nacre-Like Nanostructured Composites with Antimicrobial Properties," Langmuir 21, 11915-11921 (2005).

14) S.Deville, E. Saiz, R.K. Nalla and A.P. Tomsia, "Freezing as a Path to Build Complex Composites" Science 311, 515-518 (2006).

15) T. Naganuma, H kakisawa, A.F. Dericioglu, Y, kagawa, "Fabrication of Alumina/Polyimide Multi-Layered Material by Sputtering and Vapor Deposition Polymerization," J. Ceram. Soc. Jpn. 114, 713-715 (2006).

16) T. Kato, A. Sugawara, and N. Hosoda, "Calcium Carbonate-Organic Hybrid Materials," Avd.mater. 14, 869-877 (2002).

17) K.L. Choy, "Chemical vapor deposition of coatings," Prog. Mater. Sci. 48, 57-170 (2003).

18) Y.L. Tu, M.L. Calzada, N.J. Phillips, and S.J. Milne, "Synthesis and Electrical Characterization of Thin Films of PT and PZT Made from a Diol-Based Sol-Gel Route," J. Am Ceram. Soc. 79, 441-448 (1996).

19) M.C. Cheung, H.L.W. Chan, Q.F. Zhou, and C.L. Choy, "Characterization of barium titanate ceramic/ceramic nanocomposite films prepared by a sol-gel process," Nanostruct. Mater. 11, 837-844 (1999).

20) Metashine Technical data, Nippon Sheet Glass Co., Ltd. (2003).

21) Metals Handbook Tenth Edition, Vol. 2, ASM International (1990).

SELF ASSEMBLED FUNCTIONAL NANOSTRUCTURES AND DEVICES

Cengiz S. Ozkan
Department of Mechanical Engineering
Bourns Hall, A305
University of California, Riverside
Riverside, CA 92521

ABSTRACT
This paper reports the self assembly of functional nanostructured materials including multi-walled Carbon Nanotube-Quantum Dot (CNT-QD) heterojunctions using the Ethylene Carbodiimide Coupling procedure (EDC). Thiol stabilized ZnS capped CdSe quantum dots containing amine terminal groups (QD-NH2) were conjugated with acid treated Multi-Walled Carbon Nanotubes (MWCNT) ranging from 400 nm to 4μm in length. SEM, TEM, EDS and FTIR were used to characterize the conjugation process.

The versatile electrical properties of carbon nanotubes (CNT) make them promising candidates for nano electronic devices, especially tunneling diodes and transistors[1,2]. In previous CNT based nano-electronic devices, control over electrical properties of the devices has been limited. In addition, researchers have relied on overlapping CNTs[2] for forming junctions, which introduces local bending that could adversely affect the electrical properties of the tubes[3]. Alternatively, covalent modifications of carbon nanotubes with metal colloids[4] and semiconducting quantum dots (QDs)[5,6] have been reported for the synthesis of heterojunctions for electronics applications. These approaches were associated with carbon nanotube sidewall functionalization.

Here, the synthesis of heterojunctions with controlled conjugation of water stabilized, amine terminating, ZnS coated CdSe QDs (QD-NH2) to acid treated ends of multi walled CNTs (MWCNT) are reported. Scheme 1 illustrates the steps involved in the synthesis of the heterojunctions. Carboxyl groups were introduced to the ends of as-grown MWCNTs (Nanostructured & Amorphous materials, Inc., Los Alamos) by refluxing at 130 °C in concentrated nitric acid for 24 hours. The tips of MWCNTs, which have the highest defect sites, get oxidized first. Oxidized MWCNTs are shorter and have terminating carboxylic groups that impart a hydrophilic nature and facilitate further functionalization. ZnS capped CdSe QDs (Evident Technologies, Inc., New York) were used in the functionalization of the MWCNTs. The use of ZnS coating over the CdSe core improves the quantum yield by passivating[7] surface dangling bonds (carrier trap sites) and also, enabling them for use in biosystems[8].

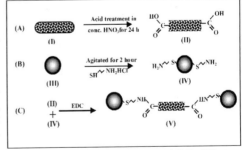

Scheme 1. Conjugation of ZnS capped CdSe QDs to MWCNTs.

To prepare water stabilized QDs (QD-NH2), ZnS capped CdSe nanocrystals were suspended in chloroform by sonication for 30 minutes. Equal volumes of 1.0 M 2-aminoethane thiol hydrochloride (AET) were added to this solution. This resulted in a two-phase mixture with the aqueous AET forming an immiscible layer above the organic chloroform-QD suspension. The mixture was stirred vigorously on a magnetic plate for 4 hours. When ZnS capped CdSe QDs were reacted with AET, the mercapto group in the thiol group in AET bonded to the Zn atoms and the amine groups rendered the QDs hydrophilic, in addition to facilitating further functionalization possibilities. The aqueous phase containing QD-NH2 was extracted in phosphate buffer saline (PBS, pH=8.5). Successive washing in PBS produced a good suspension of water stabilized QD-NH2 which were used for the synthesis of MWCNT-QD heterostructures via the two-step coupling procedure using 1-ethyl-3-(3-dimethyl-aminopropyl) carbodiimide HCl (EDC, Pierce Chemicals, Inc.) in the presence of N-hydroxysuccinimide (sulfo-NHS, Pierce Chemicals, Inc.) The EDC reaction was carried out in PBS for 8 hrs at 50°C under continuous mixing. Characterizations of the heterostructures were conducted using scanning electron microscopy (SEM), transmission electron microscopy (TEM), Fourier transform infrared spectroscopy (FTIR) and energy dispersive spectroscopy (EDS). Figure 1(A) is an SEM image of QDs conjugated at the end of a MWCNT. No sidewall functionalizations were observed. MWCNTs produce multiple carboxylic groups upon oxidation at their ends and these results in the conjugation of multiple QDs at the ends.

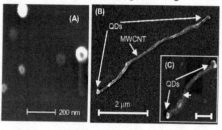

Figure 1. (A) SEM image of water soluble QDs before conjugation. (B) SEM image of QDs conjugated at the ends of a MWCNT (4μm) (C) QDs at the ends of a MWCNT (500nm).

Previous research[6] indicated sidewall QD conjugations for MWCNTs exceeding 200 nm in length. Our conjugation technique is specific that even for MWCNTs as long as 4 μm (Figure 1 (B)), QDs are observed only at the ends. Figure 1(C) is a SEM image indicating QDs conjugated at the ends of a 500 nm long MWCNT. In all the images, we have observed the absence of side wall functionalization. Further TEM (Figure 2 (A) & (B)) and EDS (Figure 2 (C)) confirms the presence of QDs at the ends of the MWCNTs.

The TEM image in Figure 2(A) shows a MWCNT with QDs at its end. Figure 2 (B) is the image of the same QD cluster at a higher magnification. Figure 2(C) shows the EDS data for the QD cluster in Figure 2 (B) using a 1.50 nm probe size. The spectra shows clear peaks corresponding to the elements of Cd, Se, Zn, and S. Notice that, Cu signal is generated from the TEM grid. The potential argument for the clusters to be Fe nano particles (from the MWCNT growth process) instead of QDs can be excluded after observing absence of the characteristic Fe peaks in EDS data. The MWCNT-QD conjugates were also characterized with FTIR, using an AgCl cell in a Bruker Equinox 55 FTIR spectrometer. Figure 3 is the FTIR spectra of oxidized MWCNTs (blue curve) and the MWCNT-QD conjugates (red curve). With plain oxidized MWCNTs, absorption peaks are observed at 1644 cm-1, 1704 cm-1 and 3403 cm-1 (peaks designated A, B, and C), which are characteristic of carboxylic and phenolic groups on acid treated MWCNTs. For the MWCNT-QD conjugates, new absorption peaks appear at 1653 cm-1, 2977 cm-1 and 3314 cm-1 (D, E, and F), which indicate the C=O stretch in amides, C-H stretch and N-H stretch in amides. C-H and N-H peaks are bigger than the amide C=O peak due to the presence of free QDs in the sample. Nevertheless, a slight blue shift of carboxylic C=O stretch to amide C=O stretch and the appearance of C-H and N-H peaks indicate the formation of covalent MWCNT-QD conjugations, via the amide bond formation.

In conclusion, a method to synthesize heterojunctions of individual MWCNTs with QDs, preferentially at the MWCNT ends has been reported. This was confirmed by SEM, TEM and EDS analysis. Due to the mild and well-controlled oxidation of the MWCNTs, conjugation occurs only at the ends for nanotubes ranging from 400 nm to 4 µm in length. The controlled conjugation process preserves the electronic properties of the MWCNTs and enables the nanoassembly of heterojunctions. They can be used as building blocks for various nanoscale electronic or optoelectronic devices and three dimensional hierarchical assemblies of multilayered systems.

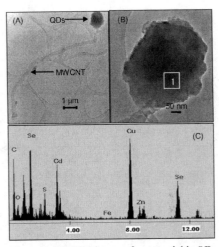

Figure 2. (A) SEM image of water soluble QDs before conjugation. (B) SEM image of QDs conjugated at the ends of a MWCNT (4µm) (C) QDs at the ends of a MWCNT (500nm).

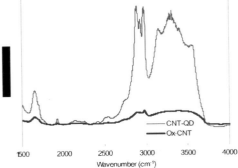

Figure 3. FTIR spectra of Oxidized MWCNTs (blue) and MWCNT-QD conjugates (red). Absorption peaks are observed at 1644 cm⁻¹, 1704 cm⁻¹ and 3403 cm⁻¹ in the FTIR spectra for oxidized tubes. New peaks develop at 1653 cm⁻¹, 2977 cm⁻¹ and 3314 cm⁻¹ in the FTIR spectra of MWCNT-QD conjugates, indicating formations of MWCNT-QD conjugates via amide bond formation.

REFERENCES
[1] (a) Tans, S. J.; Verschueren, A. R. M.; Dekker, C. Nature 1998, 393, 49. (b) Li, J.; Papadopoulos, C.; Xu, J. Nature 1999, 402, 253. (c) Yao, Z.; Postma, H. W. C.; Balents, L.; Dekker, C. Nature 1999, 402(6759), 273. (d) Ahlskog, M.;Tarkiainen, R.; Roschier, L.; Hakonen, P. Appl. Phys. Lett. 2000, 77, 4037. (e) Zhou, C. W.; Kong, J.; Yenilmez, E.; Dai, H. J. Science 2000, 290, 1552. (f) Ahlskog, M.; Hakonen, P.; Paalanen, M.; Roschier, L.; Tarkiainen, R. J. Low Temp. Phys. 2001, 124, 335. (g) Rosenblatt, S.; Yaish, Y.; Park, J.;

Gore, J.; Sazonova, V.; McEuen, P. L. Nano Lett. 2002, 2, 869.
[2]Fuhrer, M. S.; Nygard, J.; Shih, L.; Forero, M.; Yoon, Y. G.; Mazzoni, M. S. C.; Choi, H. J.; Ihm, J.; Louie, S. G.; Zettl, A.; McEuen, P. L. Science 2000, 288, 494.
[3]Dai, H. J. Surface Sci. 2002, 500, 218.
[4](a) Liu, J.; Rinzler, A. G.; Dai, H.; Hafuer, J. H.; Bradley, R. K.; Boul, P. J.; Lu, A.; Iverson, T.; Shelimov, K.; Huffman, C. B.; Rodriguez-Macias, F.; Shona, Y.-S.; Lee, T. R.; Colbart, D. T.; Smalley, R. E., Science 1998, 280, 1253. (b) Azamian, B. R.; Coleman, K.S.; Davis, J. J.; Hanson, N.; Green, M.L.H. Chem. Commun. 2000. 4, 366.
[5]Banerjee, S.; Wong S. S. Nano Lett. 2002, 2, 195.
[6]Haremza, J. M.; Hahn, M. A.; Krauss, T. D. Nano Lett. 2002, 2, 1253
[7]Hines, M. A.; Gnyotsionnest, P. J. Phys. Chem. 1996, 100, 468.
[8]Chan, W. C. W.; Nie S. M. Science 1998, 281, 2016.

COMPOSITION-STRUCTURE-PROPERTIES CORRELATION IN THE SiO$_2$-P$_2$O$_5$ SOL-GEL FILMS OBTAINED WITH DIFFERENT PRECURSORS

M. Zaharescu, L. Predoana, M. Gartner, M. Anastasescu, L. Todan, P. Osiceanu
Department of Oxide Materials Science, Institute of Physical Chemistry "Ilie Murgulescu" of the Romanian Academy, 060021 Bucharest, Romania

C. Vasiliu, C. Grigorescu, G. Pavelescu
Department of Optospintronics, National Institute of Research and Development for Optoelectronics-INOE 2000,
Bucharest-Magurele, Romania, 76900

Thin films in the SiO$_2$-P$_2$O$_5$ system are intensely studied due to their applications in microelectronics, sensing, nano-photonics, optoelectronics and as ionic conductors. Sol-gel is the most used method for preparation of such films. Previous studies established the very low reactivity of the phosphorous alkoxides and the high tendency of the phosphorous oxide to volatilize at thermal treatment. In order to identify the most appropriate precursor for obtaining layers with desired composition and properties a systematic study of the sol-gel film preparation using different phosphorous precursors was carried out. The films were deposited on glass and ITO/SiO$_2$ coated glass substrates at room temperature. To check the influence of the type of precursors on the films thermal stability and properties a post deposition annealing was performed at low temperatures in the range 150 - 200°C. The film characterization was performed using various techniques as Fourier Transform Infrared Spectroscopy (FTIR), thermogravimetric an thermodifferential analysis (DTA/TGA), X-ray Photoelectron Spectroscopy (XPS), Spectroscopic Ellipsometry (SE), Atomic Force Microscopy (AFM). The P evolution during the drying process was determined in the films obtained with P-alkoxides. Only in the films obtained with H$_3$PO$_4$ the P could be detected after thermal treatment at 200°C. This result could be explained by the fact that the phosphorous component is not chemically bonded in the matrix, but only embedded and so, it can be easily evaporated from the film during the XPS analysis.

INTRODUCTION

The SiO$_2$-P$_2$O$_5$ system has been intensively studied lately from both scientific and practical point of view. The scientific interest for the mentioned system arises from the similitude of P and Si coordination. The simple SiO$_2$-P$_2$O$_5$ oxide system is the initial component for the synthesis of multicomponent glasses with applications in many fields as: optoelectronic[1,2], generating energy such as hydrogen fuel cells and proton exchange fuel membranes[3-6].

Different applications of such materials as in microelectronics[7], nano-photonics[8], optoelectronics[2] and as ionic conductors[4-6] require thin films.

The sol-gel method enables inorganic oxides preparation at low temperatures by hydrolysis and polycondensation of the precursors in solution. This method avoids high temperature and is favourable for working with compounds with high phosphorus content, because of the high volatility of the phosphorus oxide. For this reason, sol-gel is the most used method for the preparation of the films mentioned above.

Previous studies established the very low reactivity of the phosphorous alkoxides and the high tendency of the phosphorous oxide to volatilize at thermal treatment[5,6].

The goal of this paper was the characterization of the gel and films obtained by sol-gel method starting with different precursors.

EXPERIMENTAL

Samples preparation

Phosphosilicate glasses were obtained in the 90 mol % SiO_2 - 10 mol % P_2O_5 system by sol-gel method from tetraethylorthosilicate (TEOS) as Si precursor and different P precursors as: triethylphosphate (TEP), triethylphosphite (TEPI) and phosphoric acid.

TEOS in ethanol (EtOH) was hydrolyzed with HCl/H_2O at room temperature by stirring and then the P precursor was added. The solutions were kept at room temperature in closed vessels to form gels. The wet gels were dried at 50^0C, in open vessels.

In the case of using phosphoric acid as P precursor due to the high gelation tendency, the ratio of ethanol/Σ of precursors was increased to 10 mols. Table I presents the experimental parameters used for obtaining the gels.

From the solutions prepared as presented above films were deposited on glass and ITO coated glass plates which were appropriately cleaned before the deposition. They were first dried at room temperature and then thermally treated at low temperatures (150 and 200^0C) for 15 minutes in order to avoid as much as possible the phosphorus loss from the film.

More details about the preparation procedures were given in previous works [9].

Tabel I. Experimental parameters for preparing gels in the binary system
90 molar % SiO_2 - 10 molar % P_2O_5

Sample	Precursors	Molar ratio			Reaction conditions		pH of the mixture	Time of gelling (days)
		EtOH/Σ precursors	H_2O/Σ precursors	HCl/Σ precursors	T (^0C)	Time (h)		
A	TEOS, TEPI	4	2	0,001	20	3	3	30
B	TEOS, TEP	4	2	0,001	20	3	3,5	60
C	TEOS, H_3PO_4	10	2	0,001	20	0,5	3	20

Samples characterization

Structural characterization of the samples was made via XRD and IR. IR spectra were recorded on a Perkin Elmer SPECTRUM 100 spectrometer in the range 550 - 4000 cm^{-1}. All spectra were obtained using a UATR accessory with a resolution of 8 cm^{-1}, 32 scans and a CO_2/H_2O correction.

The thermal behavior of the gels was checked by DTA/TG analysis using TGA Q 500 V5.3 Build 171 equipment.

Chemical composition of the films was examined using X-ray photoelectron Spectroscopy (XPS). XPS measurements were performed on a VG ESCA 3 MkII spectrometer, operating at a pressure of 5×10^{-9} Torr and using the non-monochromatic AlK radiation (1486.6 eV). The overall energy resolution, as determined by the full width at half maximum (FWHM) of the Au4f7/2 peak of a standard gold sample, was around 1.2 eV. In order to take into account the charging effect on the measured binding energies (BE), the spectra were calibrated using C1s line (BE = 285 eV) of the adventitious carbon on the sample surface.

Structural and optical characterization of the samples was made by Spectroscopic Ellipsometry (SE). The measurements were performed in air, in the 400 - 700 nm wavelength range at an angle of incidence of 70^0. The ellipsometric spectra have been fitted with a model consisting of three layers and six components (SiO$_2$, ITO, P$_2$O$_5$, TEP, TEPI and voids) based on Bruggemann's Effective Medium Approximation (B-EMA)[10]. The volume fractions of the components and the thickness of the layers were taken as fitting parameters. From the best fit the thickness of the films, **d**, and the volume fractions of the components were obtained (Table IV). Dielectric constants taken from literature data bases (SiO$_2$[11], ITO[12], P$_2$O$_5$[13]) have been used as references in the ellipsometric program, while TEP and TEPI were measured by a Pulfrich refractometer.

A Nomad microscope from Quesant, in the intermittent contact regime was used for the Atomic Force Microscopy images. In this regime, a vibrating tip is scanned over the surface while a feedback loop tries to maintain the vibration amplitude constant by modifying the average separation between the tip and the surface. The vibration amplitude is of the order of tens of nanometers, as well as the mean separation between the tip and the surface.

RESULTS AND DISCUSSION

Gels characterization

Starting with the initial composition and in the experimental conditions presented in the Table I, homogeneous and transparent gels were obtained. The gels were amorphous according to the XRD results (not presented here).

The IR spectra of the gels are presented in Figure 1.

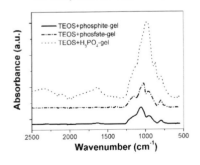

Figure 1. IR spectra of the SiO$_2$-P$_2$O$_5$ gels obtained starting with different P precursors

For all the samples the characteristic vibration bands of the Si-O-Si network inn the 1000-1200 cm^1 range could be noticed. However the intensity of the bands for the gels prepared starting with P-alkoxides is much lower than that obtained for gels with H_3PO_4 as P precursor. This could be explained by the low reactivity of the P-alkoxides that are embedded in the silica network leading to a disordered structure.

The thermal behavior of the obtained gels is summarized in the Table II, showing differences in the temperature of decomposition and in the corresponding thermal effects, depending on the P precursors used. The decomposition of alkoxidic precursors (phosphite or phosphate) occurs mainly in the 150-300^{0}C range with an exothermal effect about 260^{0}C, due to the combustion of organic residue.

If the gels are obtained starting from phosphoric acid, the main weight loss takes place below 150^{0}C, being assigned to water elimination. For all the samples water, alcohol evolution and organic combustion take place at temperatures lower than 300^{0}C. The results are shown in table II and are in agreement with the previous reported data[14].

Table II. Results of thermal analysis of the phosphosilicate gels
obtained with different P precursors

Sample	Temperature range ^{0}C	Thermal effect endo	exo	Weight loss, %	Assignment
A.	20-150	100	-	13.41	Elimination of absorbed H_2O and of EtOH
	150-300	230	260	4.93	P-precursor decomposition and combustion of organic residue
	300-500	330	-	1.58	Structural OH elimination
	500-1000	-	-	-	
	20-1000			19.92	
B.	20-150	80	-	14.60	Elimination of absorbed H_2O and of EtOH
	150-300	-	260	3.19	P-precursor decomposition and combustion of organic residue
	300-500	-	-	0.97	Structural OH elimination
	500-1000	-	-	-	
	20-1000			18.76	
C.	20-120	110	-	40.34	Elimination of absorbed H_2O and of EtOH
	120-300	-	140	6.79	Combustion of organic residue
	300-500	-	-	-	
	500-1000	-	-	-	
	20-1000			47.13	

The chemical composition of the gels prepared with different P precursors, as determined by XPS is presented in the Figures 2-4.

One may notice that samples A and B exhibit a similar behavior with the binding energies of 2p photoelectron line of Si at 104.2 eV, characteristic to SiO_2 gel and at 135.2 eV, respectively characteristic to P in combination.

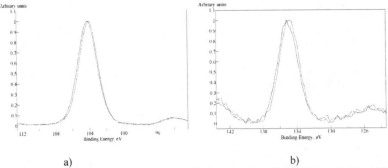

a)   b)

Figure 2. XPS spectra of the SiO_2-P_2O_5 gel obtained starting with TEPI (sample A): a) Si 2p and b) P 2p, (blue – dried at room temperature; red - thermally treated at 300^0C)

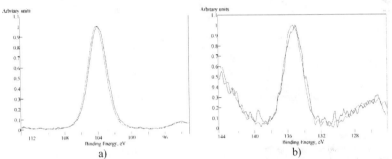

a)   b)

Figure 3. XPS spectra of the SiO_2-P_2O_5 gel obtained starting with TEP (sample B): a) Si 2p and b) P 2p, (blue – dried at room temperature; red - thermally treated at 300^0C)

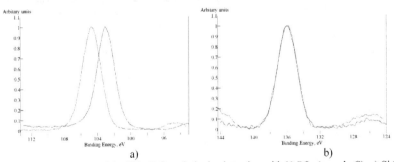

a)   b)

Figure 4. XPS spectra of the SiO_2-P_2O_5 gel obtained starting with H_3PO_4 (sample C): a) Si 2p and b) P 2p, (blue – dried at room temperature; red - thermally treated at 300^0C)

In the sample C, prepared starting with H_3PO_4 the Si 2p photoelectron line in as prepared gel is found at 103.8 eV, characteristic for the Si in a phosphosilicate structure, while after thermal treatment it shifts to a value characteristic for Si in a SiO_2 gel structure. The results could be correlated to the broken of Si-O-P bonds formed during the sol-gel process by thermal treatment.

For quantitative measurements it was measured the entire area under 2p doublets and the obtained results after spectra deconvolution are presented in the Table III.

Table III. Composition of SiO_2-P_2O_5 gels dried and thermally treated at 300^0C

Sample	Dried gels		Thermally treated at 300^0C	
	Si (atomic %)	P (atomic %)	Si (atomic %)	P (atomic %)
A	81.90	18.10	83.17	16.83
B	89.50	10.50	91.81	8.19
C	80.77	19.23	80.20	19.80
Calculated values	80.31	19.69	-	-

One may notice that sample C have a rather close composition to the nominal one. In the sample A and B a decrease in the P amount compared to the nominal one is observed, more evidenced in the later case. This fact could be correlated to the different reactivity of the used precursors. The ethylphosphate is the less reactive in the sol-gel process, leading to its embedment in the silica matrix without any chemical bonding, thus allowing an easier evaporation.

Coatings characterization

In order to compare the behavior of the gels and coatings of the same composition, deposited on different substrates (glass and ITO coated glass) the corresponding IR spectra were recorded.

The IR spectra of the glass and ITO/glass uncoated supports were also recorded (not presented here). As it was expected the crystallized ITO coatings do not present IR vibration bands, while the glass substrate presents the typical IR spectra.

The results for the as prepared gels and coatings obtained from different precursors are presented in Figure 5 (a,b,c).

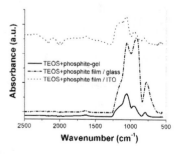

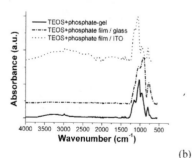

(a)   (b)

In the sample C, prepared starting with H$_3$PO$_4$ the Si 2p photoelectron line in as prepared gel is found at 103.8 eV, characteristic for the Si in a phosphosilicate structure, while after thermal treatment it shifts to a value characteristic for Si in a SiO$_2$ gel structure. The results could be correlated to the broken of Si-O-P bonds formed during the sol-gel process by thermal treatment.

For quantitative measurements it was measured the entire area under 2p doublets and the obtained results after spectra deconvolution are presented in the Table III.

Table III. Composition of SiO$_2$-P$_2$O$_5$ gels dried and thermally treated at 300^{0}C

Sample	Dried gels		Thermally treated at 300^{0}C	
	Si (atomic %)	P (atomic %)	Si (atomic %)	P (atomic %)
A	81.90	18.10	83.17	16.83
B	89.50	10.50	91.81	8.19
C	80.77	19.23	80.20	19.80
Calculated values	80.31	19.69	-	-

One may notice that sample C have a rather close composition to the nominal one. In the sample A and B a decrease in the P amount compared to the nominal one is observed, more evidenced in the later case. This fact could be correlated to the different reactivity of the used precursors. The ethylphosphate is the less reactive in the sol-gel process, leading to its embedment in the silica matrix without any chemical bonding, thus allowing an easier evaporation.

Coatings characterization

In order to compare the behavior of the gels and coatings of the same composition, deposited on different substrates (glass and ITO coated glass) the corresponding IR spectra were recorded.

The IR spectra of the glass and ITO/glass uncoated supports were also recorded (not presented here). As it was expected the crystallized ITO coatings do not present IR vibration bands, while the glass substrate presents the typical IR spectra.

The results for the as prepared gels and coatings obtained from different precursors are presented in Figure 5 (a,b,c).

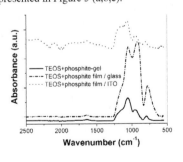

(a)   (b)

For SiO_2-P_2O_5 coatings obtained with ethylphosphate, a typical amorphous SiO_2 spectrum is noticed. By thermal treatment at 150^0C a disorder of the structure is revealed, that could be assigned to the P-precursor evaporation. By thermal treatment at 200^0C the IR spectrum became similar to the initial one.

The as deposited coating, obtained from H_3PO_4, presents the characteristic band of amorphous SiO_2 slightly shifted to lower wavenumber due to the interaction with the H_3PO_4. No considerable change is observed after thermal treatment at 150^0C, while by thermal treatment at 200^0C a structural ordering occurred.

The spectroellipsometric data are summarized in Table IV.

One may notice that the thickness of the films is different depending on the type of precursor used. The thinner films are obtained starting with H_3PO_4. By thermal treatment at 150^0C the densification of the films occurs and their thickness decrease (except the B-type samples). By further thermal treatment the thickness of the films slightly decreases for A and B samples and remains unchanged for C samples.

Due to the fact that the ethylphosphate do not hydrolize and do not condense during the sol-gel process and the ethylphosphite only partially hydrolize but do not condense, the ellipsometric spectra have been fitted taking into consideration the presence of unreacted P-precursors. The results obtained have shown much lower amount of P in samples A and B than the amount introduced in the composition. By thermal treatment the amount of P in these two samples decreases up to their complete elimination.

The evolution of the XPS spectra of the coatings obtained starting with ethylphosphate (sample B) and H_3PO_4 (Sample C), deposited on ITO coated glass substrate are presented in Figure 8 (a, b).

In the case of the films prepared starting with TEP (Fig. 8a) the characteristic P2p photoelectron line cannot be detected, meaning that phosphorus is under the XPS detection limit $(10^{-2} - 10^{-3}$ of a monolayer) rather than it has been completely (entirely) lost. In the same time, it must be emphasized that the XPS detection volume is found in the first $20 - 30$ monolayers for our investigated system. This is in good agreement with the results obtained for TEP-gel and can be explained by the fact that phosphorus component is not chemically bonded, but only embedded into the silica matrix and consequently it can be easily evaporated.

Table IV Spectroellipsometric results of the sol-gel SiO_2-P_2O_5 films obtained from different precursors deposited on ITO coated glass substrate

Sample	d(Å)	SiO₂ (%)	P₂O₅ (%)	TEP (%)	TEPI (%)	Air (%)
A	2377	92.32	-	-	7.03	0.65
A150	2127	93.73	-	-	5.74	0.53
A200	2057	99.79	-	-	0	0.21
B	2637	96.86	-	3.11	-	0.03
B150	2667	97.62	-	2.19	-	0.20
B200	2617	100	-	0	-	0
C	2120	90.90	8.95	-	-	0.15
C150	2090	90.90	8.94	-	-	0.16
C200	2080	90.90	8.94	-	-	0.16

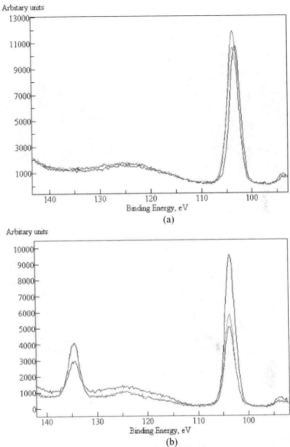

Figure 8. XPS spectra evolution of the samples B prepared starting with TEP (a) and sample C prepared starting with H₃PO₄. (b) by thermal treatment (blue – as prepared, green - thermally treated at 150⁰C, red – thermally treated at 200⁰C)

XPS spectra of the films prepared starting with H₃PO₄ presented in Figure 8 b, showed the presence of P in the as-deposited films and in the thermally treated ones. However, one may noticed some changing in the relative intensity of Si and P spectra by thermal treatment. This fact could be assigned to the changing of film morphology.

The different results obtained by SE and XPS concerning the P content of the coatings could be explained by the fact that by XPS only near surface is examined, while in SE the whole film (bulk and surface) is taken into consideration. It is possible that at the near surface the P elimination should be enhanced up to its total elimination.

AFM images are presented in the Figure 9.

TEOS +TEP (Sample B)/ITO 40x3d (a)

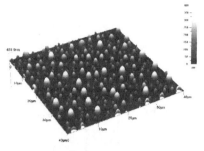

TEOS+H_3PO_4 (Sample C)/ITO 40x3d (b)

TEOS +TEP (Sample B)/glass40x3d

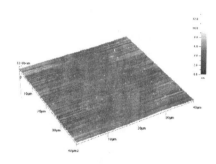

TEOS+H_3PO_4(Sample C)/glass40x3d (d)

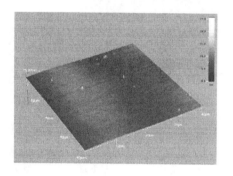

TEOS +TEP (Sample B)200/ITO 40x3d (e)

TEOS+H_3PO_4 (Sample C)200 /ITO 40x3d (f)

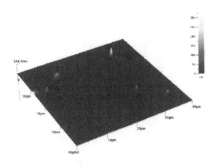 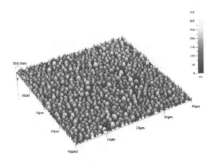

TEOS +TEP(Sample B)200 / glass   TEOS+H_3PO_4(Sample C)200/glass

Figure 9. AFM images for the SiO_2-P_2O_5 coatings obtained starting with TEP and H_3PO_4, deposited on glass and ITO coated glass substrate.

The AFM aspect of the films prepared with different precursors is very different. For the SiO_2-P_2O_5 sol gel film prepared with H_3PO_4 a globular aspect of the surface is noticed, no mater of the type of substrate used. However, the film roughness is higher on the ITO coated glass and could be correlated with the interaction of the film with the ITO layer. By thermal treatment in all cases the dispersion of the film on the substrates improved and the roughness decreased. The influence of the substrate and thermal treatment on the morphology of the SiO_2-P_2O_5 coatings is still under evaluation.

CONCLUSIONS

The influence of the type of precursors on the coatings thermal stability and properties by a post deposition annealing at temperatures of 150 and 200^0C was studied.

In the prepared films an important P loss was established, mainly when P-alkoxides were used as P precursors. The P volatilization is much lower in bulk gels.

For film preparation in the SiO_2-P_2O_5 system, special care has to be taken in order to retain the P in the film composition.

The P-alkoxide precursors are not recommended for SiO_2-P_2O_5 films preparation.

ACKNOWLEDGEMENTS

The authors express their gratitude to Dr. C. Logafatu, C. Negrila and Dr. A. Moldovan for XPS and AFM measurements. The work presented in this paper was made in the frame of National Project No. 495/15.10.2005, CE-EX type (ASMOM).

REFERENCES

[1]C. Coutier, W. Meffre, P. Jenouvrier, J. Fick, M. Audier, R. Rimet, B. Jaquier, and M. Langlet, "The effects of phosphorous on the crystallization and photoluminescence behaviour of aerosol-gel deposited SiO_2-TiO_2-Er_2O_3-P_2O_5 thin films", *Thin Solid Films*, **392**, 40-49 (2001).

[2]Repub. Korean Kongkae Taeho Kongbo KR 2001 47, 296 (Cl.G02B6/10Jun 2001)-patent

[3]A. Matsuda, T. Kanzaki, Y. Kotani, M. Tatsumisago, and T. Minami, "Proton conductivity and structure of phosphosilicate gels derived from tetraethoxysilane and phosphoric acid or triethylphosphate", *Solid State Ionics*, **139**, 113- 9 (2001).

[4]M. Aparicio, F. Damay, and L.C. Klein, "Characterization of SiO_2-P_2O_5-ZrO_2 sol-gel/Nafion composite membranes", *J.Sol-Gel.Sci.Technol.*, **26**, 1055- 59 (2003).

[5]M. Nogami, Y. Daiko, Y. Goto, Y. Usui, and T. Kasuga, "Sol-gel preparation of lost proton-conducting P_2O_5-SiO_2 glasses", *J.Sol-Gel.Sci.Technol.*, **26**, 1041- 44 (2003).

[6]Aparicio and L.C. Klein, "Sol-gel synthesis and characterization of SiO_2-P_2O_5-ZrO_2", *J.Sol-Gel.Sci.Technol.*, **28**, 199-204 (2003).

[7]B.G. Bagley, W.E. Quinn, S.A. Khan, P.Barboux, and J.M. Tarascon, "Dielectric and high Tc superconductor applications of sol-gel and modified sol-gel processing to microelectronics technology", *J. of Non-Crystalline Solids*, **121**, 454-462 (1990).

[8]B.I.Lee, Z.Cao, W.N.Sisk, J.Hudak, W.D.Samuels, and G.J. Exarhos, "Photoresponse of Tb3+ doped phosposilicate thin-films, *Mat. Res. Bul.*, **32**, 1285-92 (1997).

[9]M. Anastasescu, M. Gartner, A. Ghita, L. Predoana, L. Todan, M. Zaharescu, C. Vasiliu, C. Grigorescu, and C. Negrila, "Loss of phosporous in silica-phosphate sol-gel films", *J. Sol-Gel Sci. Technol.*, **40**, 325-333, (2006).

[10] D.A.G., Bruggeman, Berechnung verschiedener physikalischer Konstanten von heterogenen Substanzen. Ann. Phys., **24**, 636-679 (1935)

[11] E.Palik, Handbook of Optical Constants of Solids, Academic Press, Inc., NY, 1985.

[12] http://www.sungj.com/nkDatabase.jsp

[13] Kelliher John, Fundamentals of Optical Communications, (course EE/3351), Kings Colleges Physical Sciences Engineering School, University of London, 1990

[14]Y. Zhang, M.Wang, "A new method to probe the structural evolution during the heat treatment of SiO_2-P_2O_5 gel glasses", *Mat.Sci.Eng.*, **B67**, 99-101 (1999)

[15] Ph. Massiot, M.A. Centeno, I. Carrizosa, J. A. Odriozola, "Thermal evolution of sol-gel-obtained phosphosilicate solids (SiPO)", *J.Non-Cryst.Solids*, **292**, 158-166 (2001).

PREPARATION OF ZEOLITE/CARBON COMPOSITES VIA LTA ZEOLITE SYNTHESIS IN THE MACROPORES OF UNMODIFIED CARBON SUPPORTS

S.Jones[a], E.Crezee[b], P.A.Sermon[a] and S.Tennison[b]

[a] Chemistry, SBMS, University of Surrey, Guildford, Surrey, GU2 7XH, UK
[b] MAST Carbon Technology Ltd., Henley Park, Guildford, GU3 2AF, Surrey, UK

ABSTRACT
    Linde Type A (LTA) zeolite has been successfully grown in the macrostructure of carbon substrates by hydrothermal synthesis. Two unmodified carbons (i.e. MAST NOVACARB™ monolithic and willow charcoal) have been used. Within such hosts LTA crystals grow without blocking the meso or micropores of the carbon substrate and so it is expected that fluid transport through the substrate will not be hindered. If the adsorptive properties of the LTA and the carbon remain (or show synergy) then such zeolite/carbon composites could be of interest in control of a wider range of pollutants and toxic agents than previously was possible with either the zeolite or the carbon alone. In addition, it may be possible to regenerate these composites rapidly by resistive heating using the conductivity of the carbon substrate. Interestingly, calcination of the zeolite/carbon composites produces a zeolite structure that is a replica of the now-gasified carbon substrate.

INTRODUCTION
Zeolites and carbons are often used to adsorb and control air and liquid phase pollutants and toxic agents. However, there are many instances where one needs to control a combination of pollutants and in such fields composites may have advantages over sequential beds. These composites could also have advantages in delivery of oxygen-enriched air if they exhibit faster heat and mass transfer to the supported zeolites with lower pressure drops[1, 16]. Supported zeolites and zeolite membranes are of course already known[2-14], but the advantage of a carbon-hosted zeolite composite is that the substrate does not undergo dissolution in the zeolite synthesizing solutions[15]. Although it does suffer from substrate loss in oxidising atmospheres at high temperature, this can be of value in producing zeolite-only membranes on gasification of the carbon. Here the authors have considered the synthesis and properties of zeolite/carbon composites with a view to assessing these as a replacement for pelleted zeolites (where binders may interfere in an application by blocking pores)[1, 16]. The specific objective of this study was to synthesize and grow LTA zeolite crystals in the macropores and macrochannels of an unmodified *natural* and an unmodified *synthetic* carbon substrate to assess thereby the composite for suitability in control of pollutants that can damage the environment and human health.

EXPERIMENTAL
The authors could have introduced pre-prepared zeolite into the polymer precursor of synthetic MAST Carbon before carbonisation, but this would not have been applicable to the naturally-derived charcoal and could have weakened the MAST Carbon host (as could chemical modification of the carbon). It was expected that LTA zeolite crystals[17] in the macrotexture of the carbons would be too large to enter and block the host mesopores and micropores[18], which (had it occurred) the authors thought would have been detrimental for fluid transport.

Supports and their pre-treatment

Two carbon hosts with macro-channels/pores were used without prior chemical modification: (i) willow carbon (3cm x 0.5cm diameter; total surface area $24m^2g^{-1}$; see Figure 1) and (ii) MASTCarbon synthetic monolith (diameter 3cm x 0.1cm; 39 cells per cm^2; total surface area $610m^2g^{-1}$; see Figure 2). Pre-dried carbon substrates (16h at 373K) were used in all experimental work.

Synthesis of LTA zeolite in the carbon pores

LTA zeolite was synthesised using a simple hydrothermal method[17] (although the intention is to use sol-gel methods later). The composition of the LTA synthesis solution (3.165 Na_2O : Al_2O_3 : 1.926 SiO_2 : 128H_2O) was expected to give $Na_{12}[(AlO_2)_{12}(SiO_2)_{12}].27H_2O^{17}$. First 0.363g of analytical grade NaOH (Fisher) was dissolved in 40cm^3 of water and into this sodium metasilicate (Fisher, 15.48g) was introduced. Second 0.363g of NaOH was dissolved into 40cm^3 of water and sodium aluminate (Fisher, 8.26g) was introduced. Two syntheses: (i) a sodium aluminate pre-dipping and (ii) sodium metasilicate pre-dipping were used. The carbon sample was added to one of the solutions and then the other solution was introduced to give a thick gel. In a third method the carbon was dipped into the pre-formed gel. In each case the sample was vigorously shaken for 5-10min[17] and then the gel and carbon samples were aged at 373K for 12h. The carbon was removed, cooled and washed with 1dm^3 deionised water to pH 9 and then dried in an air oven for 16h at 373K.

Characterisation of the materials

LTA/carbon composites were characterised by scanning electron microscopy (SEM), X-ray diffraction (XRD) and N_2 adsorption at 77K. Specifically the morphology and composition of the composites were investigated by SEM (and associated EDX) on a Hitachi S-3200N SEM and Oxford Instruments Inca system, with a detection area of $10mm^2$ and a tungsten filament. The SEM EDX was calibrated using a cobalt standard. XRD was performed on a PANalytical X'Pert Pro MPD powder X-Ray diffractometer with monochromator. $CuK\alpha1. = 1.5405$Å radiation was used to characterise the crystalline phase synthesised and scanning range was $5°$ <2θ< $60°$. Total surface areas and pore volumes were estimated from N_2 adsorption at 77K using Micromeritics ASAP 2010 instrument.

RESULTS AND DISCUSSION

Willow carbon will have small longitudinal channels derived directly from the wood cell structure. Clearly it will also have tranverse and radial permeability (K) in channels, but this is expected[20,21] to be very much smaller than in the longitudinal direction (Figure 1a) and thus is much less important in the present preliminary work. Even though these tranverse cells are bigger (Figure 1b) they cannot be strongly interconnected and are not easily filled by precursor solutions. Micrographs of willow carbon (Figure 1) show the wood cell structure. It is predominantly made up of 2 regions: the outer region consists of small longitudinal channels running parallel to the inner region of transverse structure that appears to be of lower density 'air pockets'. These 'pockets' appear to be inaccessible to the LTA precursor gel (after hydrothermal treatment this region was shown to be devoid of LTA crystals). BET nitrogen adsorption results also suggest fluid transport is limited in this transverse material (i.e. surface area was expected to be significantly greater than was found experimentally).

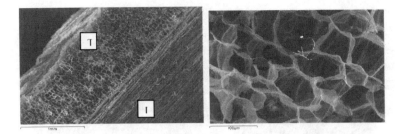

(a) willow carbon cut longitudinally (1mm scale); (b) transverse cells (100µm scale)

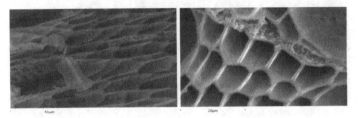

(c) longitudinal channels (40µm scale);   (d) longitudinal (20µm scale)

Figure 1. Micrographs of willow carbon

The structure of the synthetic MASTCarbon monolith comprised of two scales (see Figure 2): the visible channel structure of the monolith (a ≈700 µm^2) and the space between the particlesthat make up the walls of the monoliths (b ≈8 µm average diameter).

(a) MAST channels (1mm scale);   (b) texture of channel walls (20µm scale)

Figure 2. Micrographs of MASTCarbon monolith

Zeolite from Free Solution
SEM and XRD of the LTA zeolite crystals produced from the mixed homogeneous solution were undertaken and analysed for morphology (Figure 3a) and structure (Figure 3b) respectively. The XRD profile was consistent with that of LTA (01-073-2340, $Na_{12}Al_{12}Si_{12}O_{48}(H_2O)_{27}$); indicating that the desired LTA zeolite had been synthesised. SEM suggested that its average crystal size was 1μm.

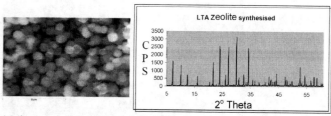

Figure 3 (a) Micrograph (8μm scale) and (b) XRD of LTA synthesised in homogeneous soln.

Zeolite Growth within Carbons
Having confirmed LTA crystallisation from free solutions, the authors now considered crystallisation from solutions within the carbons. Zeolite LTA crystals that grew in the macropores and macrochannels of the carbon samples are shown in Figure 4 (a,c: LTA/willow carbon and b,d: LTA/MASTCarbon). The crystals in both natural and synthetic carbon are the same size and shape as those formed from free solution but are smaller (at 1μm) than the expected 2-3μm[17](Figure 3a). Nonetheless they are of cubic (or dodecahedral)[17] morphology as expected from this type of hydrothermal synthesis in free solution.

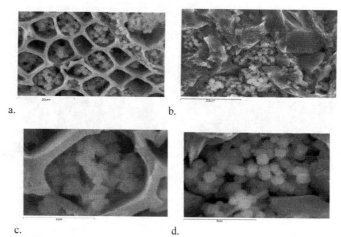

a.   b.

c.   d.
Figure 4. Micrographs of cubic 1μm LTA crystals grown in (a,c) willow carbon and (b,d) MASTCarbon (a,b: 20μm scales; c,d: 8μm scales)

These LTA crystals aggregated in the carbons did not completely obstruct the channels. Thus they are **not** expected to greatly hinder gas and liquid access to the pores of the carbon host. Consider this again in terms of N_2 adsorption at 77K as described later. Table I. gives the EDX-derived elemental composition of the LTA crystals produced in the free solution and the two carbons. The Al:Si atomic ratios are as expected of LTA zeolite (*i.e.* there should be equal amounts of Si and Al) although there appear to be elevated levels of sodium in some areas. Natural willow carbon also contained P, Ca, K and Mg.

Table I. Atom % composition of LTA crystals formed in different environments

	Free Solution	MastCarbon	Willow Carbon
Si	3.89	3.62	2.31
Al	4.01	3.79	2.13
Na	5.27	4.39	2.74
Al/Si expected	1.00	1.00	1.00
Al/Si seen	1.03	1.05	0.92

EDX is only semi-quantitative and some variation in Al:Si ratio across the sample is to be expected. * Commercial 4A showed Al/Si = 0.98

Zeolite content
Gasification of the composites (pre-dried at 373K for 16h) in air at 873K for 10h allowed the wt % of the LTA in the carbon composites to be determined gravimetrically from the residues. The LTA content of MASTCarbon sodium aluminate pre-dipped composite (5.0±0.72wt%), the sodium metasilicate pre-dipped composite (3.7±0.13wt%) and the gel dipped composite (3.4±0.48wt%) results were close to the results obtained for a gel dipped LTA/MASTCarbon composite using TGA (3.2% when heated to above 1073K). Such concentrations are good and such loadings bode well for future samples. LTA/willow composites were placed in an air furnace at 923K overnight then at 1223K for 8h. The willow pre-dipped in sodium metasilicate (9.8±2.37wt%) and the non-dipped willow (7.8±3.15wt% LTA) again had LTA levels close to that of a recently synthesised batch of gel-dipped LTA/willow samples analysed using TGA (i.e. 9.6% when heated to above 1073K). The LTA loading on willow samples appears to be consistently greater than on MASTCarbon. However, there was a variability of wt% loading on the naturally inhomogeneous willow carbon e.g. LTA on willow composite samples synthesised in the same hydrothermal gel pot was 4.65% to 10.95%. Substrate shape and density also appear to play a part in LTA adhesion (*i.e.* LTA adheres quite strongly to the outer ends of the willow carbon (Figure 5a)). While the denser MASTCarbon material is made up predominantly of micropores (0.8nm) inaccessible to the LTA gel. The macro and mesoporous sized spaces within the precursor resin are specifically engineered to allow for fluid transport to such micropores and these also allow hydrothermal gel transport into the monolith (Figure 9b). If the transverse texture is not, as assumed here, interconnected then one can understand why the LTA gel access is limited to the willow carbon longitudinal region. MASTCarbons can be engineered to allow for greater LTA loading (see Figure 5c). Nevertheless the zeolite content in these carbons is respectable for practical applications.

Figure 5a. LTA/willow (end)   b. LTA/MAST   c. LTA/MAST rod
(all 1mm scale)

The LTA weight loading on the willow samples (mainly in the longitudinal channels) is therefore greater than the MASTCarbon, but its variability is also greater since the willow charcoal pores are less homogeneous. The authors feel it is better to await a more detailed analysis of LTA levels in various carbon hosts before attributing differences in LTA loadings between MASTCarbon and Willow Carbon based composites to particular carbon characteristics (e.g. surface, chemical or textural differences).

Crystalline phase
Powder XRD profiles of carbons and composites are shown in Figure 6. These should be compared with the profile for LTA produced in free solution (see Figure 3b and reference (01-073-2340) data). The recipe used in this experiment included only sodium (with the Si/Al precursors) so should produce a 4A LTA zeolite. Potassium and calcium being larger/smaller than sodium forces the unit cell to expand or contract. This would be noticed in XRD as a slight peak shift that could be calculated using the reference cell for LTA 01-073-2340. Thus unit cell parameters were of considerable interest with regard to the LTA/willow carbon composite (since it was likely the host contained impurities such as K and Ca). The LTA/willow composite XRD scan showed peak height differences and extra peaks (see Figure 6d). Therefore in order to verify that the zeolite LTA grown within the willow channels was identical to that grown in the MASTCarbon. The unit cell parameters were calculated from peaks at approx 7,10,12.5,16,21.6,24,27 and 30 °2θ. These are shown in Table II: clearly these unit cell parameters are similar to the reference unit cell parameters.

Table II. Unit cell dimensions calculated (±0.0001Å)

LTA type	Unit cell (Å)
LTA/willow	24.575(σ = 0.02)
LTA/MAST Carbon	24.575(σ = 0.02)
Ref. cell 01-073-2340	24.610

Hence XRD of composites indicates that LTA zeolite has been successfully grown in both willow and MAST Carbon. However, there is a peak at 38°2θ in the LTA/willow composite that was absent from the LTA/MASTCarbon composite or a hydrothermally-treated willow carbon blank. It must be concluded that the LTA (or precursor) interacts such that there are two crystalline phases formed (the expected LTA and a second crystalline phase). The authors are in the process of identifying the second crystalline phase. The authors are most concerned with developing zeolite/carbon composites, but will in future for completeness determine the structure of the residual zeolite replica after carbon gasification, though this may change at the carbon gasification temperatures.

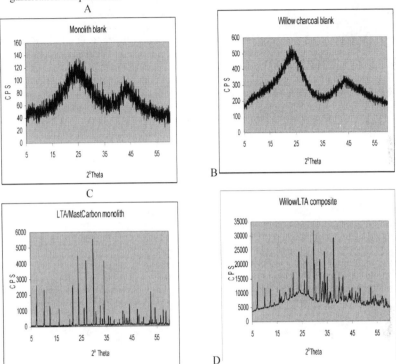

Figure 6. XRD of carbons and composites: a) MASTCarbon, b) willow charcoal, c) LTA/MASTCarbon and d) LTA/willow charcoal

N_2 adsorption

Total surface areas (S_{BET}) for the carbon hosts and composites deduced from isotherms in Figure 7 are tabulated in Table III, along with selected pore volumes (PV). LTA zeolite incorporated into the carbon host was not expected to block pores or be detrimental to the carbon functionality in any way. This is because in MASTCarbon most of the surface area is due to the micropores (<2nm), which are smaller than the LTA unit cell size (~2.4nm); its macropores are mainly useful to facilitate gas and liquid transport to its other pores.

TableIII. Surface areas $(S_{BET}\pm5m^2g^{-1})$ and pore volume $(PV\pm0.005cm^3g^{-1})$

Sample	S_{BET} (m^{2}g^{-1})	PV for pores between 1.7 and 300nm (cm^{3}g^{-1})
MASTCarbon	610.3	0.019
LTA/MAST	515.2	0.021
willow	23.7	0.019
LTA/willow	21.4	0.015

Others[25] have applied BET and BJH analysis to N$_2$ adsorption data on zeolites at 77K but the authors appreciate the limitations of these analyses.

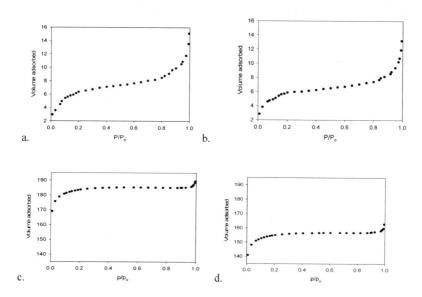

Figure 7. N$_2$ adsorption isotherms at 77K on a) willow charcoal, b) LTA/willow, c) MASTCarbon and d) LTA/MASTCarbon (where volume is in cm^3 N$_2$ (STP)/g sample)

The isotherms indicate that the total surface area and the pore volume of MASTCarbon are lowered by LTA crystallization, while in willow charcoal nitrogen adsorption is slightly less affected by the LTA incorporation (see Figures 7a-d). Interestingly triplicate measurements on commercial 4A (sigma M2135) and the LTA produced in free solution gave S_{BET} average values (after outgassing at 150°C and 250°C) of only 0.11m²/g and 6.72m²/g respectively. Therefore these are encouraging results that suggest that although LTA zeolite is aggregating in the pores and channels it is not blocking them significantly.

FURTHER WORK - ADSORPTIVE AND ION EXCHANGE

Experiments are in progress to assess (i) the ion exchange capacity and (ii) the adsorption capacity to both basic, acidic and hydrocarbon pollutants of the LTA zeolite/carbon composites described here. For (ii) a rig has been specifically designed to enable the mass flow of simple and complex gaseous mixtures to be controlled and assessed simultaneously in terms of breakthrough profiles as they pass through the composites. This will allow an assessment of these composites in purification of polluted atmospheres and waters, where there may well be competitive processes just as there are in the real environmental scenarios. It is expected that the LTA zeolite will give problems at high humidity, although this would give an excellent opportunity to assess the composites in terms of rapid regeneration using the resistivity of the carbon host.

GASIFICATION OF CARBON

Composite gasification was carried out at 873K in air to remove the carbon host and in order to quantify the % zeolite within the carbon host. This gasification produced a good zeolite replica of the initial MASTCarbon host (see Figure 8a) proving the zeolite was evenly distributed throughout the macropores of the precursor carbon substrate. This may be a route to unsupported zeolite structures and membranes. The strength of such replicas still needs to be assessed and increased if they are to be used alone. The micrographs in Figure 9a,b showed that zeolite replica of the sodium aluminate pre-dipped LTA/MASTCarbon composite after gasification indicated that much of the replica is indeed LTA zeolite, with some amorphous material (possibly unreacted sodium aluminate).

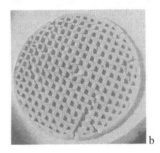

a   b

Figure 8. a) MastCarbon template and b) zeolite replica.

A similar effect was seen for all samples tested.

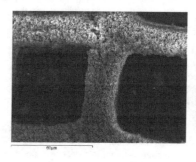 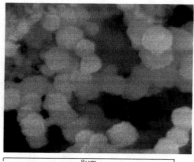

Figure 9 a) SEM of zeolite replica and b) LTA type crystals within sample Figure 9a
(after C/Au sputtering).

CONCLUSION

Many previous studies[1,15,16,19] have stressed the importance of modifying the carbon surface in order to increase the number of surface functional groups that might be essential for zeolite growth. The presence of strongly acidic surface groups is also thought to favour the anchoring of the gel on the support before crystal formation[1,16]. However, the present work suggests that the growth of LTA zeolite crystals in macrotextured carbons requires no surface modification of the carbon host. This is advantageous because it may help carbon maintain greater mechanical strength. Furthermore, chemical modifiers may promote side reactions and may weaken the binding of the LTA to the carbon support when washed out at a later stage. In addition such modifications may also increase production time and cost for the composite. Carbons, however, are produced and activated in many different ways and this (as well as the type of coating used) may be a factor determining whether surface modification is required. Others have explored carbon composites and catalysts[5,8,15,16,19,22,23,24] and seen the importance of preparative conditions on the zeolite loading achieved. It may well be that the nature (natural and synthetic), properties (chemical and textural) and pre-treatment of the carbon host are critical in defining the type and extent of LTA/carbon composite formed. It is hoped that future work will reveal this. Interestingly raising the hydrothermal aging time does not appear to influence LTA zeolite crystal size (i.e. a hydrothermal synthesis of 2 days (rather than 12 hours) still produced crystals of 1μm). The hydrothermal route chosen appears successful in producing well distributed clusters of LTA crystals inside the carbon macrotexture. These are spread evenly throughout the carbon structure without blocking the micropores of the carbon hosts. In summary a novel rapid route to zeolite/carbon composites is reported here. The authors soon expect to describe the unusual synergy-driven purification and pollution control properties for these elegant materials (and possibly other potential uses of the zeolite replicas of the carbon templates).

ACKNOWLEDGEMENTS:
The authors thank EPSRC and MASTCarbon for support of SJ via a studentship. At UniS they also thank L.Courtney (for SEM), C. Burt (for replica SEM), P.Slater (for helpful discussions) and M.A.Howard (for help with MS/rig design).

REFERENCES:
[1]A. Berenguer-Murcia, J. Garcia-Martinez, D. Cazorla-Amoros, A. Linares-Solano and A.B. Fuertes, Silicalite-1 membranes supported on porous carbon discs, *Micropor. Mesopor. Mater.*, **59**, 147-159 (2003).
[2]X. Chen, W. Yang, J. Liu and L Lin, Synthesis of zeolite NaA membranes with high permeance under microwave radiation on Mesoporous-layer-modified macroporous substrates for gas separation., *J. Membr. Sci.*, **255**, 201-211 (2005)
[3]X. Xu, W. Yang, J.Liu and L.Lin, Synthesis and perfection evaluation of NaA zeolite membrane., *Sep. Purif. Tech.*, **25**, 475-485 [1] (2001)
[4]A. Huang, Y.S. Lin and W. Yang, Synthesis and properties of A-type zeolite membranes by secondary growth method with vacuum seeding., *J. Membr. Sci.*, **245**, 41-51 (2004)
[5]V. Engstrom, B. Mihailova, J. Hedlund, A Holmgren and J.Sterte, The effect of seed size on the growth of silicalite-1 films on gold surfaces., *Micropor. Mesopor. Mater.*, **38**, 51-60 (2000)
[6]T. Masuda, H. Hara, M. Kouno, H. Kinoshita and K. Hashimoto, Preparation of an A-type zeolite film on the surface of an alumina ceramic filter., *Micropor. Mater.*, **3**, 565-571 (1995)
[7]X. Gu, J. Dong, T.M. Nenoff and D.E. Ozokwelu, Separation of p-xylene from multicomponent vapour mixtures using tubular MFI zeolite membranes., *J. Membr. Sci.*, **280**, 624-633 (2006)
[8]K. Okada, Y. Kameshima, C.D. Madhusoodana and R.N.Das, Preparation of zeolite-coated cordierite honeycombs prepared by an in situ crystallization method., *Sci. & Tech. Adv. Mater.*, **5**, 479-484 (2004)
[9]S. Yamazaki and K. Tsutsumi, Synthesis of an A-type zeolite membrane on silicon oxide film-silicon, quartz plate and quartz fiber filter., *Micropor. Mater.*, **4**, 205-212 (1995)
[10]M. Sathupunya, E. Gulari and S. Wongkasemjit, Na-A (LTA) zeolite synthesis directly from alumatrane and silatrane by sol-gel microwave techniques., *J. Europ. Cer. Soc.*, **23** 1293-1303 (2003)
[11]Y. Li, X. Wang, S. Zhang and D. Wang, Preparation and characterization of A-type zeolite/Si O2 molecular sieving membranes., *Trans. Nonferrour. Met. Soc. China*, **13** (1), 55-59 (2003)
[12]S. Miachon, E. Landrivon, M. Aouine, Y.Sun, I. Kumakiri, Y. Li, O. Pachtova Prokopova, N. Guilhaume, A. Giroir-Fendler, H. Mozzanega and J.A. Dalmon, Nanocomposite MFI-alumina membranes via pore-plugging synthesis, Preparation and morphological characterisation., *J. Membr. Sci.*, **281**, 228-238 (2006)
[13]B. N. Nair, T. Yamaguchi, T. Okubo, H. Suematsu, K. Keizer and S.I. Nakao, Sol-gel synthesis of molecular sieving silica membranes., *J. Membr. Sci.*, **135**, 237-243 (1997)
[14]X. Xu, M. Cheng, W. Yang and L.Lin, Synthesis and gas permeation properties of silicalite-1 zeolite membrane., *Sci. in China (Series B)*, **41** (3), 325-330 (1998)
[15]X. Zhang, T. Wang, H.Liu and K.L. Yeung, Preparation of composite carbon-zeolite membranes using a simple method., *J. Mater. Sci.*, **39**, 5603-5605 (2004)
[16]J. Garcia-Martinez, D. Cazorla-Amoros, A. Linares-Solano and Y.S.Lin, Synthesis and characterization of MFI-type zeolites supported on carbon materials., *Micropor. Mesopor. Mater.*, **42**, 255-268 (2001)

[17]R.W. Thompson, M.J. Huber, K.C. Franklin and (H. Robson Editor), LTA Linde Type A, *Verified Syntheses of Zeolitic Materials.,* Elsevier Science **B.V,** 179-181 (2001)

[18]S.R. Tennison, Phenolic-resin-derived activated carbons., *Applied Catalysis A: General,* **173,** 289-311(1998)

[19]J.Ozaki, K.Takahashi, M.Sato and A,Oya, Preparation of ZSM-5 nanoparticles supported on carbon substrate., *Carbon,* **44,** 1243-1249 (2006)

[20]Y.Taniguchi and S.Nishio, High-frequency electric-power vacuum-drying of wood 6. Gas-permeability during vacuum drying and drying characteristics of various species., *Moguzai Gakkaishi* **39** (3), 277-283 (1993)

[21]T.Lihra, A.Cloutier and S.Y. Zhang, Longitudinal and transverse permeability of balsam fir wetwood and normal heartwood., *Wood Fiber Sci.* **32** (2), 164-178 (2000)

[22]R.Hubart, J.Altafulla, A.Rives and C.Scott, Characterization and HDS activities of mixed Fe-Mo sulphides supported on alumina and carbon., *J. Fuel* **86** (5-6), 743-749 (2007)

[23]C.Zollfrank, R.Kladny, H.Sieber and P.Greil, Biomorphous SiOC/C-ceramic composites from chemically modified wood templates., *J.Europ.Cer.Soc.* **24,** 479-487 (2004)

[24]G.Onyestyak, J.Valyon and K.Papp, Novel biomorphous zeolite/carbon composite having honeycomb structure., *Mater. Sci & Eng. A* **412,** 48-52 (2005)

[25]. A.Derkowski, W.Franus, H.Waniak-Nowicka and A.Czimerova, Textural properties vs. CEC and EGME retention of Na-X zeolite prepared from fly ash at room temperature., *Int. J. Miner. Process.* **82,** 57-68 (2007)

RAPID INSPECTION OF CARBON NANOTUBE QUALITY

S.A. Hooker and R. Geiss
Materials Reliability Division
National Institute of Standards and Technology
Boulder, CO 80305

R. Schilt
Department of Applied Mathematics
University of Colorado
Boulder, CO 80305

A. Kar
Department of Chemical and Biological Engineering
Rensselaer Polytechnic Institute
Troy, NY

ABSTRACT

Carbon nanotubes have unique properties of interest for applications in aerospace, electronics, and biotechnology. However, the properties of different batches of carbon nanotubes can vary considerably depending on chemical purity and the nanotube types present (*e.g.*, diameter and chirality distribution). Distinguishing the constituents of each nanotube batch is challenging, with many different techniques used in concert. Thermogravimetric analysis (TGA) provides one measure of nanotube purity by assessing the material's thermal stability (*i.e.*, how it oxidizes with temperature). Unfortunately, however, TGA analysis requires a relatively large specimen for each measurement (several milligrams), making it inappropriate for rapid screening of incoming materials. Moreover, the measurement provides only an average purity for the analyzed sample, and variability can occur on a much finer level. As many applications will utilize only a small quantity of nanotubes, new approaches are needed to assess variability for a much smaller specimen size. This paper describes a new analysis method that uses a quartz crystal as a miniature microbalance for determining mass changes at elevated temperature. Thin nanotube coatings are spray deposited onto the crystals, and shifts in a crystal's resonance frequency are directly correlated with changes in coating mass during heating due to volatilization of different carbon species. By monitoring the response of the crystal at one or more temperatures, different nanotube specimens can be directly compared. This paper demonstrates concept feasibility by comparing quartz crystal results with conventional TGA analysis and discusses methods for applying the technique in process and quality control settings.

INTRODUCTION

Carbon nanotubes exhibit electrical, mechanical, and thermal properties that make them attractive for numerous applications, including multifunctional composites, biomedical devices, drug delivery, and next-generation electronics, among many others[1]. However, their chemical and structural characteristics can have a strong influence on material behavior[2], making it important for product developers to be able to rapidly inspect new materials post-synthesis in order to optimize product performance and ensure quality control. In addition, certain chemical impurities that result from the various material synthesis processes can affect dispersion stability, with even a small change in concentration leading to a significant effect on processability. Finally, purification to remove such impurities can introduce nano- and macro-scale defects which can alter the interface between the nanotubes and other materials (*e.g.*, a polymer matrix), again affecting the ability to uniformly apply the nanotubes in a given application. These issues are further complicated by the fact that most synthesis routes for carbon nanotube manufacture do not produce a homogeneous material, but instead generate a mixture of nanotube types (*i.e.*, different diameters, lengths, and chiral angles), along with potentially large quantities of carbonaceous and metallic impurities. As a result of this variability, nanotube characterization remains a key challenge for material manufacturers, application developers, and regulatory agencies investigating potential health and safety risks.

Thermogravimetric analysis (TGA) is commonly employed to help assess nanotube purity[3,4,5]. This technique analyzes changes in the weight of a specimen in relation to changes in temperature under given atmospheric conditions. In the case of carbon nanotubes, an oxygen-containing atmosphere (*i.e.*, air) is typically used, allowing the different forms of carbon present to oxidize with increasing temperature. This oxidation results in a net weight loss over time. As weight loss curves only provide the onset temperature for oxidation, it is typical to examine the derivative of the weight loss versus temperature, with the maximum value corresponding to the average oxidation temperature. When this oxidation temperature is low, the material is generally considered to be less pure, allowing one to compare materials from batch to batch to obtain a relative measure of carbon purity[6]. The maximum temperature for the measurement is selected to be sufficiently high as to allow all of the carbonaceous material to oxidize, leaving behind only residual metal. This metal typically consists of catalytic particles from the synthesis reactions, as well as impurities introduced during purification or mixing.

TGA analysis does provide a qualitative measure of the carbonaceous species present, as well as the quantity of residual metal. However, it should be noted that conventional TGA instruments determine an average purity across a relatively large quantity of bulk nanotubes (*i.e.*, several milligrams).   Because many applications will utilize these materials in much smaller quantities, a technique capable of quickly sampling and assessing homogeneity in a small specimen is needed. This paper describes just such a new measurement approach, based on a quartz crystal microbalance (QCM) platform. A QCM is a sensing device capable of measuring small mass changes in real-time by monitoring shifts in the resonance frequency of a thin quartz crystal (*i.e.*, the frequency of minimum impedance). When the crystal is oscillated, its resonance frequency will decrease proportionally as a mass is applied to the crystal surface.   QCM sensitivity is extremely high, allowing one to measure changes as small as a few nanograms[7]. At present, QCMs are used to monitor film thickness, detect the presence of toxic gases, and measure the progression of molecular interactions such as moisture uptake in paints[8].

In addition to these more conventional applications, advanced measurement systems based on QCM technology are under development. For example, a new QCM-based calorimeter has been reported which measures mass changes, heat flows, and viscoelastic damping in thin films[9]. The QCM is thermally coupled to a heat sink through a Peltier thermopile used as a heat flow sensor. The device is placed in an adiabatic thermal environment for stability, and the QCM is coated with a thin film of the desired test material. As the film undergoes exothermic reactions (*e.g.*, on exposure to humid air), heat flow is generated and detected as a voltage change at the thermopile. At the same time, the change in mass associated with water adsorption is detected at the QCM. By combining data, complete reaction processes can be described.

Here, we utilize an alternative method for analyzing exothermic reactions in carbon nanotubes. A small quantity of bulk carbon nanotubes is placed directly in the center of a QCM, shifting the resonance frequency of the crystal in direct proportion to the applied coating mass. The crystal is then heated to one or more temperatures, resulting in a weight loss due to volatilization of the nanotubes. This weight loss is registered by a corresponding shift in the resonance frequency of the crystal, bringing it closer to the original value. By tracking frequency shifts with temperature, specific points on the TGA curve can be analyzed. The quantity of tubes sampled by this technique is considerably less than that needed for TGA measurement (i.e., a few micrograms as compared to a few milligrams). As a result, this approach can bring to light variations within the material on a much finer scale than possible with conventional analytical instruments. In addition, because of its simplicity and the small quantity of material required for each analysis, this method is particularly well suited for rapid sampling and inspection of a new material in order to quantify its homogeneity.

To demonstrate the validity of this approach, results from elevated temperature QCM testing are compared with TGA measurements for a series of nanotube specimens. Three specific temperatures along the carbon volatilization curve were selected for QCM analysis, and the mass percentages after heating were compared to illustrate the degree of homogeneity in the material. Recommended standard practices for TGA characterization were followed[10] to maximize repeatability of analytical results.

EXPERIMENTAL PROCEDURE

Several grams of single-walled carbon nanotubes were received from a commercial supplier, all of which were produced in a single manufacturing run[11]. No purification processes were employed to remove amorphous carbon or metallic impurities, providing a real-world specimen with which to evaluate the sensitivity of the two techniques to non-nanotube constituents. Scanning electron microscopy (SEM) was used to characterize the material upon receipt, examining specifically the size and quantity of the metal particles present and the degree to which an amorphous carbon coating encased these particles.

A series of thermogravimetric analyses were performed on the material, including testing of over twenty five randomly sampled specimens (2-4 milligrams per measurement). To the maximum extent possible, different regions of the as-received bulk material were sampled to minimize any effects due to settling and differentiation of the different constituents over time. Weight loss profiles were obtained in an air atmosphere from near ambient temperature (30 °C) to a maximum temperature of 800 °C to ensure all of the carbon-containing constituents oxidized. The remaining weight percentage of the sample at 800 °C was recorded and referred to as the residual metal content (M_r). It should be noted that the final temperature was sufficiently high as to either partially or fully oxidize the residual metal catalyst remaining in the sample; therefore, this value is slightly higher than the actual metal content in the as-received material. A linear heating rate of 5 °C/minute was employed, as suggested by Arepalli, et al.[6] The derivative of the TGA weight loss curve with temperature (dm/dT) was used to determine the oxidation temperature of each sample (i.e., the maximum of the derivate, denoted T_o). In certain specimens, multiple peaks were observed in the derivative curve. These were individually determined and referred to as oxidation temperature 1, oxidation temperature 2, etc. (e.g., T_{o1},

T_{o2}). As previously mentioned, the oxidation temperature, as determined by TGA, is used by many researchers as a measure of the material's thermal stability, with higher oxidation temperatures typically associated with purer, less-defective carbon nanotubes[12]. To assess the homogeneity of the overall bulk material, the standard deviations of both the oxidation temperature and the residual mass (σ_T and σ_M) were determined. Because the initial mass used for each TGA measurement varied slightly from sample to sample, a coefficient of variance (c_v) was calculated for each parameter, indicating the % of the mean represented by the standard deviation. The material was considered to be statistically invariable when c_v was less than 2 %.

For the elevated temperature QCM measurements, a stable dispersion was first prepared by mixing approximately 2 milligrams of the bulk carbon nanotubes with approximately fifty milliliters of trichloromethane (i.e., chloroform). It should be that because of the "fluffy" nature of the carbon nanotube material, precise weighing at these quantities was not practical; as a result, the concentration reported is only approximate. The mixture was agitated ultrasonically for 45 minutes to break up the agglomerates, creating a relatively stable nanotube suspension in the chloroform. This suspension remained stable for several hours, allowing sufficient time to prepare coatings from the material. Coatings were deposited by spraying the dispersion onto masked QCMs with a small-volume spray gun.

The resonant frequency of each QCM was determined before and after the coating was applied using a standard laboratory impedance analyzer. The QCM devices used for this study were obtained commercially and possessed a resonance frequency of 10 MHz prior to coating. Upon applying the coating, the resonance frequency decreased proportionally with the applied mass, as estimated by the Sauerbrey equation[7]:

$$\Delta f = [-2 * f_0^2 * \Delta m] / [A * (\rho_q * \mu_q)^{1/2}] \qquad (1)$$

where Δf = the shift in resonance frequency (e.g., due to applied coating or on heating)
 f_0 = the resonance frequency of the uncoated QCM
 Δm = the change in mass resulting in the corresponding resonance frequency shift
 A = the active area of the quartz resonator (i.e., the electrode area)
 ρ_q = the density of quartz, 2.648 g/cm^3
 μ_q = the shear modulus of quartz, 2.947 x 10^{11} g/cm-s^2

The coated crystals were then heated to temperatures of 375 °C, 400 °C, and 425 °C using a small muffle furnace, with a heating rate of 10 °C/minute, a dwell time at temperature of 10 minutes, and a programmed cooling rate of 10 °C/minute. Achieving this cooling rate with the muffle furnace was difficult as temperature decreased, and the actual cooling rate could not be easily determined. The temperatures examined were based on initial results from the TGA analyses, in which the first oxidation temperature was near 375 °C. The resonance frequency of each QCM was re-measured after the device had cooled to room temperature, and the resulting shifts in frequency were used to determine the changes in mass due to heating, as previously described. On heating, the coating mass decreased due to volatilization of a portion of the carbon-containing components, resulting in an increase in the resonance frequency of the QCM. Figure 1 shows typical coated QCM devices, illustrating the difference in mass change when crystals were heated to 375 °C, 400 °C, and 425 °C. Seventy two QCM crystals were characterized before and after coating and before and after heating (i.e., 24 tests per temperature). The means, standard deviations, and coefficients of variance were calculated for

375 °C, 400 °C, and 425 °C, and the QCM means were compared with the average results (at each temperature) from the TGA data.

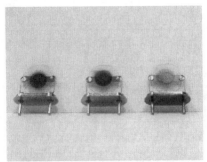

Figure 1. QCM devices coated with carbon nanotubes (*i.e.*, black coating covering the gold electrode in the center of each crystal). From left to right: the crystals have been heated to temperatures of 375 °C, 400 °C, and 425 °C, with significant mass loss occurring at 425 °C. In this latter case, the underlying gold electrode is clearly visible.

RESULTS

SEM analysis of the as-received carbon nanotube material indicated the presence of a significant quantity of impurity particles. These particles were determined by Energy Dispersive Spectroscopy (EDS) to be comprised of nickel and yttrium, likely residual catalyst materials from the synthesis process. Estimation of particle size using the image analysis capabilities of the SEM revealed that the particles had an average diameter of approximately 80 nanometers. Where carbon nanotubes were present, the materials were highly bundled, resulting in long "ropes" with no visible isolated nanotubes. Both the metallic particles and the nanotube bundles were encased in amorphous carbon, as shown in Figure 2. Because of the high surface area associated with the particulates and the extent to which the materials appear coated, it was anticipated that the amorphous carbon content of the material was relatively high. High contents of amorphous carbon significantly reduce the thermal stability of the overall material, resulting in oxidation at a much lower temperature. In addition, the presence of a large concentration of metallic particles (such as was evident for this material) can catalyze oxidation reactions, further reducing the material's thermal stability.

These observations were confirmed through TGA analysis, which revealed a relatively low onset oxidation temperature (~325 °C), as well as two distinct oxidation peaks in the derivative curve, as shown in Figure 3. The two peaks in the derivative curve (368.87 °C and 406.04 °C) occurred at significantly lower temperatures than typically observed in purified specimens (*i.e.*, 500-600 °C[12]), confirming the relatively low content of single-walled carbon nanotubes in the bulk material. In addition, the weight percentage remaining post-analysis was over 40 %, indicating that the material contained a high metal content.

As previously discussed, twenty-five specimens, randomly sampled from the original bulk material, were analyzed by TGA, resulting in average properties of T_{o1} of 369.8 ± 2.4 °C,

T_{o2} of 403.1 ± 7.2 °C, and M_r of 41.9 ± 1.2 %. Coefficients of variation were 1.39 %, 0.64 %, and 2.78 % respectively. These results indicate that the bulk material is relatively homogeneous, with only slight variability from specimen to specimen, as evidenced by the uniformity of the curves in Figure 4. The predominant source of this variability was the difference in residual metal content, which ranged from 40.52 % to 45.08 % over the course of the 25 experiments. These values represent oxidized metal, and the extent of oxidation can depend on particle size. As variations in particle diameter were observed via SEM, the recorded variability in metal content may attributable in part to differences in oxidation.

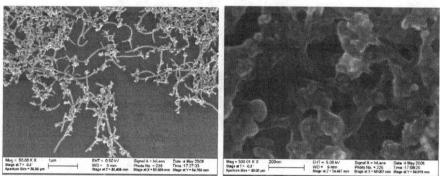

Figure 2. SEM micrographs of the as-received carbon nanotube material. The left image shows the presence of large nanotube bundles (possibly ropes), as well as a significant quantity of metallic particles. The right image shows that the particles and ropes are encased in amorphous carbon.

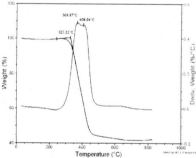

Figure 3. TGA data showing oxidation of the bulk carbon nanotube material with temperature. The deflection point in the weight loss curve (left; 327.32 °C) reflects the initial temperature at which the amorphous carbon begins to oxidize. Peaks in the derivative curve (right; 368.87 °C and 406.04 °C) represent oxidation of higher order carbons, likely carbon nanotubes.

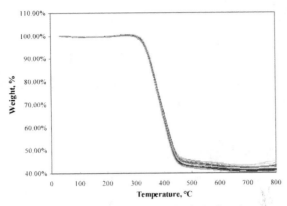

Figure 4. Comparison of 25 TGA measurements for the bulk carbon nanotube material.

As anticipated, mass changes in the carbon nanotube material were also observed on heating using the elevated temperature QCM technique. Characteristic changes in the resonance frequency of the quartz crystal were observed after application of the carbon nanotube coating and after heating of the material in air. The addition of mass due to the applied coating decreased the resonance frequency, while volatilization of material during heating increased the resonance frequency (*i.e.*, due to mass loss). Because only a fraction of the applied coating was affected by heating, a difference remained between the final resonance frequency and the initial resonance frequency (*i.e.*, for the uncoated QCM). This difference can be directly correlated to the mass remaining after heating to each specific temperature.

Figure 5 shows data for a typical crystal heated to 400 °C. The resonance frequency of the QCM was initially 9.9889 MHz. Application of the carbon nanotube coating decreased this frequency to 9.9756 MHz. From this data, the Sauerbrey equation indicates that the applied coating mass was 22.09 micrograms. Heating the crystal to 400 °C in air increased the resonance frequency to 9.9799 MHz. Here, a decrease in coating mass of 7.14 micrograms was calculated. These results indicate that 67.67 % of the initial carbon nanotube mass remained after heating this particular coating to 400 °C. This result is quite similar to the average value determined by TGA at this temperature (*i.e.*, 65.45 %).

However, despite the relative similarity of results for this particular coating to the results obtained by TGA, a comparison of elevated temperature QCM data for over 70 specimens of the bulk carbon nanotube material indicated a substantial degree of variability not observed in the TGA measurements. As shown in Table 1, while the average data for the two techniques are similar, the coefficients of variance differ substantially. For example, after heating to 375 °C the average mass remaining was 75.88 ± 16.26 % using the QCM method, with a coefficient of variance of 21.42 %. At the same temperature, TGA data indicated an average remaining mass of 77.90 ± 0.64 %, with a coefficient of variance of only 0.82 %. At all three temperatures analyzed, the average values were quite close between the two techniques; however, variability was several orders of magnitude greater for the QCM measurements. This difference can be observed visually in the histograms in Figures 6 and 7 for QCM and TGA data respectively.

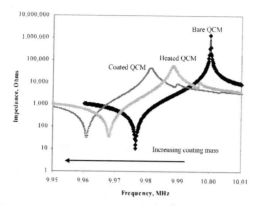

Figure 5. Impedance data for typical QCM coated and heated to 400 °C. The resonance frequency (f_0) is defined as the frequency of minimum impedance.

Table 1. Comparison of TGA and QCM results at the three examined temperatures.

	375 °C		400 °C		425 °C	
	TGA	QCM	TGA	QCM	TGA	QCM
Average	77.90 %	75.88 %	65.45 %	65.62 %	53.38 %	49.77 %
Std Dev.	0.64 %	16.26 %	0.96 %	14.02 %	1.12 %	14.55 %
Coeff. Var.	0.82 %	21.42 %	1.47 %	21.37 %	2.10 %	29.24 %

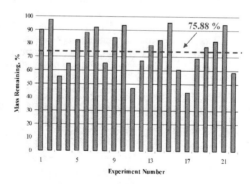

Figure 6. Histogram showing the degree of variability in thermal stability of 22 carbon nanotube coatings heated to 375 °C using the elevated temperature QCM method. The data for mass remaining ranged from 43.53 % to 97.37 %, with an average of 75.88 %.

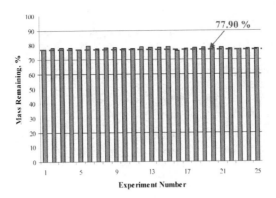

Figure 7. Histogram showing the degree of variability in thermal stability of 25 carbon nanotube specimens heated to 375 °C during TGA measurements. The data for mass remaining ranged from 76.25 % o 79.39 %, with an average of 77.90 %.

DISCUSSION

Several significant differences exist between the TGA and QCM approaches, the most notable of which is the amount of material investigated by the technique. Approximately 2-4 milligrams of bulk carbon nanotubes were analyzed per TGA run compared to 10-20 micrograms per QCM measurement. All QCM trials (~70) combined did not require as much material as one TGA measurement. As a result, the data obtained per TGA trial represents an average purity for a relatively large quantity of bulk carbon nanotubes. In contrast, each QCM trial represents the purity of only a small fraction of that same material. At the level probed with the QCM technique, variations in the size and quantity of residual metal particles, the extent of encapsulation of these particles with amorphous carbon, and the degree of macroscopic bundling of the carbon nanotubes into "ropes" can lead to substantial differences in thermal stability.

The aggregate effect of these differences can appear relatively minor, as evidenced by the apparent homogeneity observed in the repeated TGA analyses. However, when one considers how the material will be used in applications such as those in microelectronics, biotechnology, and gas detection, even small variations in material composition can have a profound effect on electrical and thermal behavior. For example, the role of metallic impurities becomes pronounced at these levels, due in part to the relatively large size of these particles (*i.e.*, average of 80 nanometers), their ability to catalyze (*i.e.*, accelerate) certain surface reactions, and their interference with electrical performance. Additionally, even subtle variations in metal content, bundling, or encapsulation can have a profound influence on the ease with which the material can be processed, either into dispersions, films, coatings, or composites. Even if sufficient properties can be achieved in the final product (*e.g.*, structural reinforcement of polymer composites), the inability to create a uniform distribution may lead to the use of higher load fractions of nanotubes, often at a significant cost penalty. These issues illuminate the need to not only analyze the overall homogeneity of a given bulk specimen, but also to assess the distribution of its constituents at a level important for product manufacturability.

This paper describes one proposed approach for screening carbon nanotube materials. Statistical comparison of the average QCM data (against average data from TGA analyses) confirmed the validity of the method as an indicator of material quality (see Table 2). Although considerable variability was observed from coating to coating, the average thermal stability was statistically the same as that observed by TGA. SEM analysis (Figure 2) confirmed the variability determined by QCM. Metallic particles represented a large fraction of the bulk material and varied in both size and distribution. Amorphous carbon was evident throughout the sample, creating agglomerated, bundled ropes that likely contained a distribution of nanotube diameters and lengths. Finally, these ropes varied in both size and degree of isolation from the metal particles. These observations clearly support the data acquired through numerous QCM trials. Each coating likely possessed subtle differences in composition leading to not so subtle differences in thermal stability.

Table 2. One sample t-test comparing QCM data to a hypothetical mean (*i.e.*, TGA average).

	375 °C	400 °C	425 °C
Number of Trials	22	23	22
Mean	75.88	65.62	49.77
Standard Deviation	16.26	14.02	14.55
Standard Error of Mean	3.47	2.92	3.10
Coefficient of Variance (%)	21.42%	21.37%	29.24%
Variance	264.31	196.60	211.82
Hypothetical Mean (from TGA)	77.90	65.45	53.38
Calculated t	-0.58	0.06	-1.16
Critical t value (95% confidence)	2.08	2.07	2.08

However, additional investigation is warranted to develop best practices for using the QCM approach for carbon nanotube analysis. Although statistical analysis indicated no differences between the TGA and QCM measurements, minor differences did exist in the time and extent of thermal exposure of the carbon nanotube material. In the case of the TGA measurements, a heating rate of 5 °C/minute was used, with no holds at temperature. In the case of the QCM measurements, a heating rate of 10 °C/minute was used, with a 10 minute soak at the maximum temperature. Moreover, TGA data were acquired during heating, whereas QCM data were acquired after the crystal had cooled to ambient temperature. While such differences are not extensive, they may account for the slight variations in the average thermal stability recorded for each technique.

In addition to differences in thermal exposure, the study described herein relied on spray deposited films from non-optimized dispersions using chloroform as the solvent. While the nanotube suspensions appeared visibly stable, non-uniform deposition was observed for many films, as shown in Figure 8. In certain instances, droplets were noted, likely due to rapid solvent evaporation at the nozzle of the spray gun. In other cases, agglomerated metal particles were clearly visible, some in excess of 5 micrometers in diameter. At the temperatures used for this study, oxidation of these particles is unlikely. As the particles contribute to the overall mass of the coating, their relative stability at these temperatures can bias the mass change data resulting in a decrease in mass loss at a specific temperature. Conversely, such particles can potentially dislodge from the coating during normal handling and testing of the QCM, resulting in an increase in mass loss not related to a thermal effect. While it is desirable for any characterization technique to be insensitive to dispersion conditions, SEM analysis of the applied coatings suggests that optimizing the dispersion for improved spraying will reduce the uncertainties associated with the measurement.

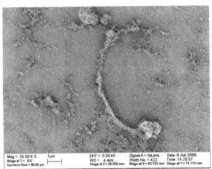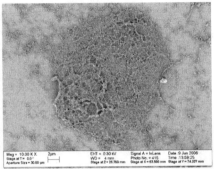

Figure 8. SEM micrographs of different regions of a carbon nanotube film. The image at left shows several large agglomerates resulting from incomplete dispersion. The image at right illustrates denser regions likely formed as droplets at the spray nozzle.

CONCLUSIONS

This paper describes a new measurement approach for determining the purity and homogeneity of small specimens of bulk carbon nanotubes. The method utilizes a quartz crystal microbalance for thermal analysis of a thin nanotube coating. Validation of this approach was achieved by comparing results with those obtained by conventional TGA analysis. All specimens were derived from a single batch of as-produced carbon nanotubes, without additional purification. TGA data indicated that the raw material contained large fractions of both amorphous carbon and residual metal catalyst. Despite the lack of purity (*i.e.*, with respect to single walled carbon nanotubes), repeated TGA measurements indicated that the material was relatively homogeneous, with only slight variability from specimen to specimen. The predominant source of this variability was the metal content. Comparison of average QCM data at 375 °C, 400 °C, and 425 °C with average TGA results at those temperatures indicated no statistical difference between the two techniques. However, QCM analysis of over 70 specimens illustrated a substantial degree of variability not observed by TGA. SEM analysis confirmed this variability. Metal particles varied in size and distribution; highly bundled nanotube ropes were present; and the entire specimen was encased in a thick amorphous carbon coating. These results illustrate that the elevated temperature QCM method can serve as an effective screening tool for assessing the quality of carbon nanotube materials, bringing to light variability at a level not previously possible with conventional analytical instrumentation.

REFERENCES

[1] W.A. deHeer. "Nanotubes and the Pursuit of Applications," *Materials Research Society Bulletin*, 281-5 (2004).

[2] L. Qingwen. Y.H.Z. Jin, and L. Zhongfan, "Dependence of the Formation of Carbon Nanotubes on the Chemical Structures of Hydrocarbons." 8th International Conference on Electronic Materials, IUMRS-ICEM, Xi'an, China, June 10-14, 2002.

[3] B.P. Ramesh, W.J. Blau, P.K. Tyagi, D.S. Misra, N. Ali, J. Gracio, G. Cabral, and E. Titus, "Thermogravimetric Analysis of Cobalt-Filled Carbon Nanotubes Deposited by Chemical Vapour Deposition." *Thin Solid Films*, 494, 128-32 (2006).

[4] G.S.B McKee and K.S. Vecchio, "Thermogravimetric Analysis of Synthesis Variation Effects on CVD Generated Multiwalled Carbon Nanotubes" *J.Phys.Chem. B*, 110, 1179-86 (2006).

[5] B.J. Landi, C.D. Cress, C.M. Evans, and R.P. Raffaelle. "Thermal Oxidation Profiling of Single-Walled Carbon Nanotubes," *Chem. Mater.*, 17, 6819-34 (2005).

[6] S. Arepalli, P. Nikolaev, O. Gorelik, V.G. Hadjiev, W. Holmes, B. Files, and L. Yowell, "Protocol for the Evaluation of Single-Wall Carbon Nanotube Material Quality," *Carbon*, 42 1783-91 (2004).

[7] D.S. Ballantine, R.M. White, S.J. Martin, A.J. Ricco, E.T. Zellers, G.C. Frye, and H. Wohtjen, *Acoustic Wave Sensors*, San Diego: Academic Press, 1997.

[8] R.F. Schmitt, J.W. Allen, J.F. Vetelino, J. Parks, and C. Zhang, "Bulk Acoustic Wave Modes in Quartz for Sensing Measurand-Induced Mechanical and Electrical Property Changes," *Sensors and Actuators B*, 76, 95-102 (2001).

[9] A.L. Smith. "Mass and Heat Flow Measurement Sensor," US Patent 6,106,149 (2000).

[10] See http://www.msel.nist.gov/Nanotube2/Carbon Nanotubes Guide.htm.

[11] Carbolex, Broomall, PA, see www.carbolex.com. *** Certain commercial equipment, instruments, or materials are identified in this paper to foster understanding. Such identification does not imply recommendation or endorsement by the National Institute of Standards and Technology, nor does it imply that the materials or equipment identified are necessarily the best available for the purpose.

[12] P. Hous, C. Liu, Y. Tong, S. Xu, M. Liu, and H. Cheng, "Purification of Single-Walled Carbon Nanotubes Synthesized by the Hydrogen Arc-Discharge Method," *J.Mater.Res.*, 16, 2526-9 (2001).

CONTINUOUS PRODUCTION AND HARVESTING OF INORGANIC-CERAMIC NANOPARTICLES

S.A.E. Abdulla[a], P.A. Sermon[a,*], M.Worsley[a] and I.R. Collins[b]

[a]Chemistry, SBMS, University of Surrey, Guildford, Surrey, GU2 7XH. UK
[b]BP. Chertsey Road, Sunbury-on-Thames., TW16 7LN, UK.

ABSTRACT

The authors demonstrate here that the continuous synthesis of $BaSO_4$, $Ca_3(PO_4)_2$, $LaCoO_3$, $BaCeO_3$ etc can be achieved to give a range of particle sizes and morphologies in water-in-TX100/cyclohexane/2-methyl-2-propanol microemulsions and can then be continuously harvested by cooling the microemulsion to 268 K. Surprisingly, the nanoparticles then appear in the *organic-rich* upper phase.

INTRODUCTION

Nanomaterials can be nm sized building blocks for a whole range of applications, e.g. synthetic bone, bone cements, pigments and ceramics provided that their preparation can be scaled up. The last ten years has seen steady development in the uses of microemulsions in the preparation of metal[1], semiconductor[1] phosphate[2], mixed oxide[3,4] (e.g. manganate[4] or ferrite[4]) and ceramic[5] nanoparticles of controlled size, composition, structure and morphology. Here the authors consider how this may be scaled up using four examples: $BaSO_4$, tricalcium phosphate/hydroxyapatite, $BaCeO_3$ and $LaCoO_3$.

Microemulsions are of interest in materials preparation as the reactions that lead to particle formation of a characteristic size and morphology[6] occur within dynamic droplets of nm (colloidal) size in which reactant concentrations define the size of solid particles formed. The microemulsion consists of three components: water, oil and surfactant (and possibly a co-surfactant). At certain compositions the system is optically clear and spontaneously formed. Generally water-in-oil microemulsions have the water phase dispersed in the hydrocarbon phase[7], but stabilised by the added surfactant.[8] The properties of microemulsions has been explained[9] in terms of the continual coalescence/re-separation of droplets.

The novel aim of this work was to *continuously* synthesise and harvest nanoparticles of $BaSO_4$ (where nanoribbons[13] have been continuously harvested), tricalcium phosphate/hydroxyapatite, $BaCeO_3$ (recently prepared by a microemulsion route[12] as 5-50nm nanoparticles) and $LaCoO_3$ by mixing separate water-in-oil microemulsions containing the reactive ions (e.g. Ba^{2+} and $(SO_4)^{2-}$ [10,11]) and then thermally cycling these around the phase inversion temperature. The nanoparticles formed were to be compared with those produced in free solution.

EXPERIMENTAL

Decane ($\rho = 0.73$ g.cm^{-3}), dodecane ($\rho = 0.75$ g.cm^{-3}), isooctane (2,2,4-trimethylpentane; $\rho = 0.692$ g.cm^{-3}), *p*-xylene (1,4-dimethylbenzene; $\rho = 0.86$ g.cm^{-3}), 2-propanol ($\rho = 0.785$ g.cm^{-3}), cyclohexane ($\rho = 0.78$ g.cm^{-3}; Fisher Scientific) and 2-methyl-2propanol ($\rho = 0.78$ g.cm^{-3}) were used as hydrocarbons. Ethanol ($\rho = 0.789$ g.cm^{-3}) was used with distilled water. The following surface-active agents have been used: (i) SP10 (Sisterna; HLB2 sucrose ester of stearate/palmitate (10% monoester)), (ii) AOT (sodium bis-(2-ethylhexylsulfosuccinate; sodium dioctyl sulfosuccinate; RMM = 444.55), (iii) $C_{12}EO_4$ or Brij30 (poly(oxyethylene-4-dodecyl ether), RMM = 362.56; density = 0.95 g.cm^{-3}), (iv) DDAB (didodecyldimethylammonium bromide; RMM = 462.65) and (v) Triton X100 or octyl phenol ethoxylate (RMM = 625; ρ = 1.07 g.cm^{-3}). In the aqueous phases the reagents used were $BaCl_2$, Na_2SO_4 (Fisher Chemicals), Na_3PO_4 (BDH), $Ca(NO_3)_2$ (BDH), $Ba(NO_3)_2$ (Sigma Aldrich), $Ce(NO_3)_3$ (Acrōs Organics), $La(NO_3)_3$, $Co(NO_3)_3$, $CuCl_2.2H_2O$ and ammonium oxalate monohydrate (RRM = 142.11; Sigma Aldrich). Unless otherwise specified all reagents were of analytical grade and obtained from Aldrich.

Densities, conductivities (Wayneker bridge (Automatic LCR 7330) and FTIR (Perkin Elmer 2000) were measured on various faces. Harvested nanoparticles were characterised here by transmission electron microscopy (TEM Philips CM200) using holey carbon grids.

Preparations in Bulk Free Solution

(a) $BaSO_4$: Using $BaCl_2$ and Na_2SO_4 (2:1) in ethanol-water (25:75) $BaSO_4$ was prepared by mixing 10 cm^3 of $BaCl_2$ solution (1.2214 g/dm^3, 0.005M) in ethanol water 25:75 and 5 cm^3 of a Na_2SO_4 solution (0.7102 g/dm^3, 0.005M).

(b) $Ca_3(PO_4)_2$/hydroxyapatite: Calcium phosphate in ethanol water 25:75; $Ca_3(PO_4)_2$ (2:3 mol ratio) [Na_3PO_4 and $Ca(NO)_2$] was prepared by mixing 10 cm^3 sodium Na_3PO_4 solution (32.69 g/dm^3, 0.086M) in ethanol+water (25:75) with 10 cm^3 $Ca(NO_3)_2$ solution (21.17 g/dm^3, 0.129M) in ethanol+water (25:75). Calcium phosphate with 1% copper doping was prepared as above except that 0.16 g $CuCl_2.2H_2O$ (1% with relation to molar Ca^{2+} concentration) was added to the $Ca(NO)_2$ (21.17 g/dm^3, 0.129M) predissolved in water.

(c) $BaCeO_3$ and $LaCoO_3$ were prepared by dissolving the nitrates of Ba and Ce (1:1) (or nitrates of La and Co) in water and then adding excess ammonium oxalate solution with agitation. The product was a mixed oxalate that was washed well with water, dried and calcined to 873 K.

Microemulsion Preparations of $BaSO_4$

(a) $BaSO_4$ in AOT/isooctane/water: 0.25 cm^3 of 0.01M $BaCl_2$ solution (2.4 g/dm^3) and 0.25 cm^3 of 0.01M (1.42 g/dm^3) Na_2SO_4 solution were added drop-wise separately to 2 cm^3 0.1M AOT (45 g/dm^3 in isoctane) each under vigorous stirring. The two microemulsions were combined to form $BaSO_4$ microemulsion. $BaSO_4$ in DDAB/dodecane/water and $C_{12}EO_4$/decane/water microemulsions were prepared as described for AOT. Surfactant concentration used was 46.3 g/dm^3 in dodecane for DDAB and 36.3 g/dm^3 in decane for $C_{12}EO_4$ surfactant. $BaSO_4$ in HLB2/xylene/propanol was prepared using 0.2 cm^3 of each $BaCl_2$ solution (0.01M, 2.4 g/dm^3) and $NaSO_4$ solution were added dropwise to 2 cm^3 SE HLB 2 (10 g/dm^3 in xylene) separately. The initial suspension was transformed to a transparent microemulsion by adding 2 cm^3 2-propanol to each emulsion.

(b) TX100/cyclohexane/2-methyl-2-propanol/water: 0.07 cm^3 of an aqueous solution of 0.1M BaCl$_2$ (24.4 g/dm^3) was added dropwise to 5 cm^3 of 0.2M TX100 (387.6 g/dm^3) (in cyclohexane and 2-methyl-2-propanol 10:1 ratio). 0.07 cm^3 of an aqueous 0.1M Na$_2$SO$_4$ (14.2 g/dm^3) solution was added dropwise to another 5 cm^{3}of 0.2M TX100 in cyclohexane and 2-methyl-2-propanol (10:1). Mixing these formed water-in-oil microemulsions. On cooling to 268 K two layers were again formed. Densities of two layers were measured (upper layer: 0.78 g.cm^{-3}; bottom layer: 0.84 g.cm^{-3}). Since the densities of both 2-methyl-2propanol and cyclohexane are 0.78 g.cm^{-3}, the water and surfactant were assumed to have been in the bottom layer.

All BaSO$_4$ microemulsions formed were colourless and transparent. Conductivities (see Table I) suggested that these microemulsions were water-in-oil. The presence of water in the bottom layer after lowering the temperature of the TX100 microemulsions was proved by both conductance and IR. Conductivity of BaSO$_4$ focussed TX100 microemulsion (after cooling to 268 K) the bottom layer was approximately 67 times higher than the upper layer (0.105 mS and 6.700 mS for upper and bottom layers respectively). IR (see Figure 1) showed that the intensity of the 3400 cm^{-1} H-O was 25 times higher in the lower than in upper layers. There was a strong C-H bend at 2922 - 2852 cm^{-1} in both layers and another strong C-H bend at 1450 cm^{-1} at upper layer and a weak one in the bottom layer (i.e it was 0.3 intensity ratio of the upper phase value (see Figure 1a,b)). Weak peaks of TX-100 were found in the bottom layer at 1524 cm^{-1} and 1512 cm^{-1} (see Figure 1).

Table I. Conductivities of microemulsions used in BaSO$_4$ preparation

Surfactant used	HLB 2	AOT	C$_{12}$EO$_4$	DDAB	TX100
PH	7	6.8	6	6	7.2
Conductivity/mS	0.17	0.11	0.16	0.91	3.1

Other TX100 Microemulsion-based Preparations

(a) Ca$_3$(PO$_4$)$_2$: The above TX100 microemulsion route was repeated using 0.1M of each Ca(NO$_3$)$_2$ (23.6 g/dm^3) and Na$_3$PO$_4$ (14.2 g/dm^3) (1:1). Densities for both upper layer (0.79 g.cm^{-3}) and bottom layer (0.89 g.cm^{-3}) were measured. Here again most water was in the bottom layer (see Figure 1).

(b) LaCoO$_3$: Here three microemulsions were produced: (i) using 0.07 cm^{3}of 0.1M La(NO$_3$)$_3$ (32 g/dm^3), (ii) 0.1M Co(NO$_3$)$_3$ (29 g/dm^3) was added dropwise to 5 cm^3 of 0.2M TX100 (in cyclohexane and 2-methyl-2-propanol 10:1 ratio) and (iii) 0.14 cm^3 of 0.1M NH$_4$OH was added dropwise to 10 cm^3 of 0.2M TX100 (in cyclohexane and 2-methyl-2-propanol 10:1 ratio). The three microemulsions were mixed together and then cooled at 268 K. Again two layers were formed. The densities of the upper (0.77 g.cm^{-3}) and bottom (0.80 g.cm^{-3}) layers were measured. FTIR again (see Figure 1) suggests water was mainly in the bottom layer.

(c) BaCeO$_3$: This preparation used the same method as LaCoO$_3$ except that 0.1M Ba(NO$_3$)$_2$ (26 g.dm^{-3}) and 0.1M Ce(NO$_3$)$_3$ (43.4 g.dm^{-3}) were used. Densities for upper and bottom layers were found to be 0.79 g.cm^{-3} and 0.84 g.cm^{-3} respectively. Again FTIR suggested (see Figure 1) that water was mainly in the bottom layer.

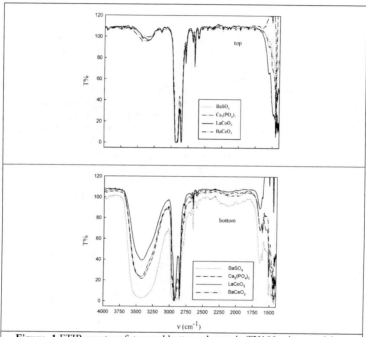

Figure. 1 FTIR spectra of top and bottom phases in TX100 microemulsions (after phase inversion at 268 K) when these were used for BaSO$_4$, Ca$_3$(PO$_4$)$_2$, LaCoO$_3$ and BaCeO$_3$ preparation

RESULTS AND DISCUSSION

Similar conductivity and FTIR evidence was seen with the TX100 microemulsions route to Ca$_3$(PO$_4$)$_2$ (see Figure 1). The conductivity of bottom layer (i.e. 7 mS) was much higher than that of the upper layer (0.12 mS) (see Table III). Furthermore IR again showed the same water (O-H) stretch at 3500 cm^{-1} where the upper:bottom intensity ratio was 0.23 upper to bottom layers. The bottom:upper intensity ratio of the C-H stretch at 1450 cm^{-1} was found to be 0.57. Again similar data were available from the LaCoO$_3$ and BaCeO$_3$ TX100 microemulsions (see Table II) after phase inversion at 268 K.

BaSO₄

Figure 2 shows a micrograph of the $BaSO_4$ 10-20nm nanoparticles formed in a free aqueous + ethanolic solution. These are clearly aggregated, but show a uniformity of primary

Table II. FTIR evidence of phase-separated TX100 microemulsions used for $LaCO_3$ and $BaCeO_3$ preparation

Stretch/cm⁻¹	LaCoO₃/cm⁻¹	%T	BaCeO₃/cm⁻¹	%T
O-H	3424 bottom	2.425	3442 bottom	3.9
	3357 upper		3357 upper	
C-H	2928-2853 bottom	similar	2926-2854 bottom	Similar
	2932-2852 upper	"	2922-2852 upper	"
C-H	1450 bottom	0.1	1450 bottom	0.31
	1451 upper		1451 upper	

Table III. Conductivities of $Ca_3(PO_4)_2$, $BaCeO_3$ and $LaCoO_3$ emulsion phases before and after phase separation

Microemulsion	BaSO₄/m S	Ca₃(PO₄)₂ / mS	BaCeO₃ / mS	LaCoO₃ / mS
Before phase inversion	3.000	4.000	1.300	3.300
After phase inversion	0.104 upper	0.115 upper	0.100 upper	0.095 upper
	6.700 bottom	7.400 bottom	5.600 bottom	8.100 bottom

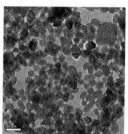

Figure 2. BaSO$_4$ nanoparticles produced in free ethanol:H$_2$O solution (scale=50nm)

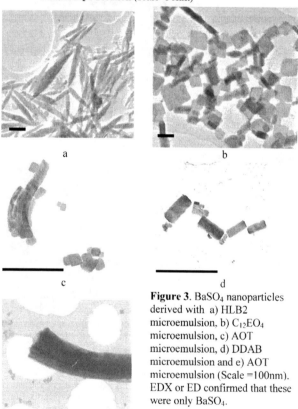

Figure 3. BaSO$_4$ nanoparticles derived with a) HLB2 microemulsion, b) C$_{12}$EO$_4$ microemulsion, c) AOT microemulsion, d) DDAB microemulsion and e) AOT microemulsion (Scale =100nm). EDX or ED confirmed that these were only BaSO$_4$.

Figure 3a-e shows TEM micrographs of BaSO$_4$ of very different morphologies formed in various transparent microemulsions. HLB2 (see Figure 3a) produced baguette particles (12-28nm x 125-300nm), C$_{12}$EO$_4$ (see Figure 3b) baguettes and thin platelets (25-60nm), AOT (see Figures 3c, 2e) platelets (18-37nm) and nanoribbons ($\geq$1 μm) and DDAB (see Figure 3d) cylindrical bundles of nanoribbons (11-60nm in length and 7-26nm width). Elongated particles with AOT are mentioned in the literature.[14,15] Ellipsoidal particles have been reported before for ZnS with Triton X100 in the presence of cosurfactant n-butanol.[16] EELS confirmed that these were BaSO$_4$ are not artefacts. However, all of those BaSO$_4$ nanoparticles were very difficult to harvest. Hence more attention was now paid to TX100 based microemulsions, where phase separation could be based on phase inversion temperatures.

BaSO$_4$ from TX100 microemulsions after Phase Inversion

Figure 4 shows that a large number of BaSO$_4$ nanoparticles were found extensively in the TX100 microemulsion upper phase (50%) after thermal separation. This suggests that the phase inversion method has delivered a usable method of nanoparticle harvesting.

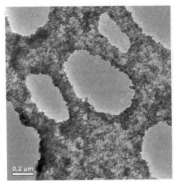

Figure 4. BaSO$_4$ nanoparticles (11-29nm) produced in upper layer.

This upper phase was water-depleted but produced a high density of harvestable nanoparticles similar to those found in free ethanol-water solutions (see Figure 2).

Ca$_3$(PO$_4$)$_2$ Results and Discussion

Gelatinous Ca$_3$(PO$_4$)$_2$ formed in free solution contained 25-62nm fibrons (see Figure 5a). Densities and conductivities (see Table III) of both upper and bottom layers of the TX100 microemulsion containing Ca(NO$_3$)$_2$ and Na$_3$PO$_4$ after phase inversion at 268 K proved that the particles were again in the upper organic layer (see Figure 4b). TEM shows that dispersed 27-107nm cubic and 80-213nm hexagonal Ca$_3$(PO$_4$)$_2$ nanoparticles are formed and harvested from the microemulsion. This is promising for injectable bone.

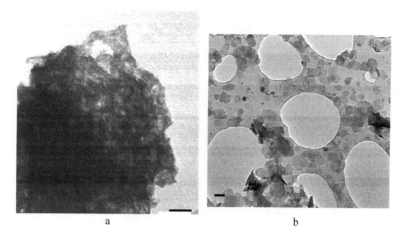

a   b

Figure 5. Ca₃(PO₄)₂ a) formed in free solution, b) in TX100 microemulsion cubic and hexagonal (scales=50 nm).

BaCeO₃ And LaCoO₃ Results and Discussion

    Figures 6a and b shows micrographs of BaCeO₃ and LaCoO₃ at 873 K formed by free bulk solution; rods were much larger than nonoparticles. Figures 7a and 7b clearly show that for both LaCoO₃ and BaCeO₃ very highly dispersed nanoparticles and nanochains were harvested in the upper phases of TX100 microemulsions on phase segregation at 268 K. Since the bottom layers had conductivities 8.1 mS and 5.6 mS for LaCoO₃ and BaCeO₃ respectively (see Table III) that were 85 and 56 times higher than corresponding upper layers. the upper layer was again thought to be organic rich. A comparison of Figures 6 and 7 looks very promising for ceramic production nano-particles.

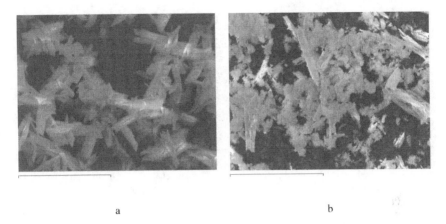

a   b

Figure 6 SEM of a) BaCeO₃ particles and b) LaCoO₃ formed in free solution (scale 30μm)

CONCLUSIONS

There is much interest in the design of materials using nanoparticles (e.g. $Ca_3(PO_4)_2$-hydroxyapatite [17,18] with and without doping [19], $BaSO_4$ [20,21,22], $LaCoO_3$ [23] and $BaCeO_3$ [24,25] and nanocomposites for a whole range of applications e.g. from synthetic bone [26] to ceramics [27].

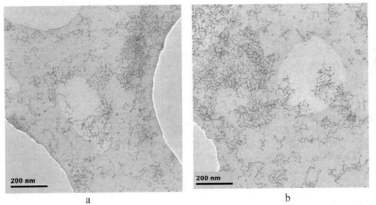

a   b

Figure 7 TEM of a) LaCoO₃ in a TX100 microemulsion after phase inversion nanoparticles (size 10-21nm) and nanothreads (size 43-108nm) were formed and b) BaCeO₃ in a TX100 microemulsion after phase inversion nanparticles (size 10-20nm) and nanothreads (size 40-121nm) were formed.

This paper shows that the continuous synthesis of well dispersed nanoparticles of $BaSO_4$, $Ca_3(PO_4)_2$, $LaCoO_3$ and $BaCeO_3$ etc can be achieved using water-in-TX100/cyclohexane/2-methyl-2-propanol microemulsions with harvesting after phase separation of an organic-rich top phase on cooling to 268 K. The benefits over free solution routes is most clear for $LaCoO_3$ and $BaCeO_3$ and this bodes well for ceramic production.

A range of particle sizes and morphologies (particles, platelets, ribbons, threads and baguettes (as previously seen[22])) can be produced for inclusion in products in a range of applications. The different morphologies of particles is determined by the relative rate of growth along different crystallographic directions. Mann[28,29] reported that faces perpendicular to the fast direction of growth have smaller surface areas and slow growing faces. Therefore fast growth along one specific axis produces a needle-shaped morphology, whereas platletlikes were produced while growth along two directions. $BaCeO_3$ and $LaCoO_3$ nanoparticles formed are primary nuclei which are involved in the growth of the nanochains.[29,30] The formation of $BaSO_4$ elongated nanoribbons was related to interweaving of surfactant bound to the long axis in adjacent nanochains (see Figure 2e).[30]

ACKNOWLEDGEMENTS
The authors thank EPSRC and BP/Viaton for funding for SA via a studentship, L. Courtney (UniS) for SEM and A. Reynolds (Brunel University) for some TEM micrographs.

REFERENCES
[1] J. H.Clint, I. R.Collins, J.A.Williams, B.H.Robinson, T.F.Towey, P.Cajean and A. Khanlodhi *Farad. Discuss.* **95**,219-233,(1993)
[2] M.Cao, X.Y.He, X.L.Wu and C.W.Hu, *Nanotechnology*, **16**,2129-2133,(2005)
[3] V.Uskokovic, D.Makovec and M.Drofenik *Curr.Res.Adv.Mater.Processes Mater.Sci.Forum* **494**,155-160,(2005)
[4] V.Uskokovic and M.Drofenik *Prog.Adv.Mater.Processes Materials Science, Forum*, **453-454**,225-230,(2004)
[5] C.Pithan, Y.Shiratori, A.Magrez, S.B.Mi, J.Dornseiffer and R.Waser *J.Ceram.Soc. Japan* **114(1335)**,995-1000,(2006)
[6] L. Qi, J.Ma, H.Cheng and Z.Zhao, *Coll. Surf.* **108A**,117-126,(1996)
[7] B.Niemann, F.Rauscher, D.Adityaawarman, A.Voigt and K.Sundmacher, *Chem. Eng. Proc.* **45**,917-935,(2006)
[8] T.Nishimi and C.A.Miller *Langmuir* **16**,9233-9241,(2000)
[9] S.Clark, D.I.Fletcher and Y.Xilin *Langmuir* **6**,1301-1309,(1990)
[10] J.D.Hopwood and S.Mann *Chem.Mater.* **9**,1819-1828,(1997)
[11] N.I.Ivanova, D.S.Rudelev, B.D.Summ and A.A.Chalykh *Coll.J.* **63**,714-717,(2001)
[12] J.Cai, K.Laubernds, F.S.Galasso, S.L.Suib, J.Liu, X.F.Shen, H.R.Kunz and J.M.Fenton *J.Amer.Ceram.Soc.* **88**,2729-2735,(2005)
[13] P.A.Sermon, N.Mason McLellan and I.R.Collins *Cryst.Eng.Comm.* **6**,469-473, (2004)
[14] M.P.Pileni and J.Tanori, *Adv.Mater.* **7**, 862-864,(1995)
[15] M. Li and S. Mann, *Langmuir* **16**,7088-7094,(2000)
[16] T.Charinpanitkul, A.Chanagul and J.Dutta, *Sci.Tech. Adv. Mat.* **6**,266-271,(2005)
[17] S.C.Lee, H.W.Choi, H.J.Lee, K.J.Kim, J.H.Chang, S.Y.Kim, J.Choi, K.S.Oh and Y.K.Jeong *J.Mater.Chem.* **17**,174-180,(2007)

[18]N.Pramanik, P.Bhargava, S.Alam and P.Pramanik *Polymer Compos.* **27**,633-641,(2006); A.Monkawa, T.Ikoma, S.Yunoki, K.Ohta and J.Tanaka *Mater.Lett.* **60**,3647-3650,(2006); Y.M.Mao, D.X.Li, H.S.Fan, X.D.Li, Z.W.Gu and X.D.Zhang *Mater.Lett.* **61**,59-62,(2007)
[19]Z.Evis *J.Ceram.Soc.Japan* **114**,1001-1004,(2006)
[20]G.G.Chen, G.S.Luo, J.H.Xu and J.D.Wang *Powd.Technol.* **153**,90-94,(2005); J.K.Liu, Q.S.Wu and Y.P.Ding *Chem.Res.Chinese Univ.* **21**,243-245,(2005)
[21]M.H.Qu, Y.Z.Wang, Y.Liu, X.G.Ge, D.Y.Wang and C.Wang *J.Polymer Sci.* **102**,564-570,(2006)
[22]H. Bala, W. Fu, J. Zhao, X. Ding, Y. Jiang, K. Yu and Z. Wang, *Colloids and Surfaces* **252A**,129-134,(2005)
[23]X.L.Cui, Q.H.Yang and X.X.Fu *J.Rare Earths* **21(Suppl S)**,124-126,(2003); M.Chen, Y.W.Wang and X.M.Zheng *Chin.J.Inorg.Chem.* **19**,1145-1149,(2003)
[24]D.W.Lee, J.H.Won and K.B.Shim, *Mater.Lett.* **57**,3346-3351,(2003)
[25]K. Ouzaouit, A. Benlhachemi, H. Benyaich, J. P. Dallas, S. Villian, J. A. Musso and J. R. Gavarri, *M. J. Condensed Mater.* **7**,94-97,(2006)
[26]J. A. Wihmhurst, R. A. Brooks and N. Rushton *J. Bone Joint Surgery (British)* **83B**,278-282,(2001)
[27]M. F. M. Zawrah and N. M. Khalil *Ceramics Intn.* **27**,309-314,(2001)
[28]S.Mann, *Angew. Chem. Int. Ed.* **39**,3392-3406,(2000)
[29]S. Sadasivan, D. Khushalani and S. Mann, *Chem. Mater.* **17**,2765-2770,(2005)
[30]G. D. Rees, R. Evans-Gowing, S. J. Hammond and B. H. Robinson, *Langmuir* **15**,1993-2002,(2002)

PREPARATION AND CHARACTERIZATION OF SUBMICROMETER-SIZED ZINC OXIDE

M. Bitenc and Z. Crnjak Orel*
National Institute of Chemistry
Hajdrihova 19, SI-1000 Ljubljana, Slovenia

ABSTRACT
The preparation of nano- and submicrometer-sized one-dimensional (1-D) zinc oxides (ZnO) was performed from zinc nitrate by precipitation with urea in two types of solvents (water and water/polyol mixtures). The influence of different polyols (ethylene glycol, diethylene glycol and tetraethylene glycol) on the size of formed particles (length and diameter) is presented for the first time. The influence of the concentration of precursors, solvents and the ratio of water/polyol, temperature, pH and time of synthesis was correlated with the size and the morphology of obtained particles. In all cases crystalline ZnO was synthesized in the form of hexagonal bipods. The morphological and crystalline properties of the synthesized samples were characterized by SEM, IR, and XRD.

INTRODUCTION
Nano- and submicrometer structured materials, especially the preparation of semiconductor materials, present one of the fast growing fields in science. ZnO presents one of the most attractive semiconductors and is becoming one of the very important materials due to its unique properties of near-UV emission (1), electric conductivity and optical transparency. It shows potential application in catalysis, optoelectronic devices, sensors, and photovoltaic. Some of one-dimensional (1-D) ZnO nano structured morphologies are wires (2), needles (3), tubes (4), columns (5), tetra pods (6), helices (7).

The solution phase approach presents the low cost preparation method and has recently attracted a lot of interest due to low temperature growth (85-95°C). It is much more appropriate for large-scale synthesis, especially nowadays, since the consumption of energy is very low. ZnO nanoparticles were prepared with chemical precipitation method with controllable morphology via solution route (8). Unfortunately, with this method either much different morphology was obtained or the distribution range of diameter was wide. For that reason, the preparation of ZnO nano structures still presents a great challenge under low temperature growth.

In the last years, polyol method was used for the synthesis of different inorganic compounds as well as pure metals (9-12). Polyol acts simultaneously as solvent, reducing agent and a medium for preventing particle growth. Due to the mild reducing power, polyols are unable to reduce a cation of electropositive metals such as zinc (9). In literature, the homogenous alkalization of metal salt solution by hydrolysis of urea was widely explored for the preparation of metal oxide (13). At our work, we combined the method of homogenous precipitation by using zinc nitrate and urea in solvent that was the mixture of water/polyol. All experiments were done under constant mixing at low temperature (90°C). The obtained materials were characterized by different techniques.

In this paper solution phase synthesis of ZnO is described. The influence of added alcohol or polyols, which serve as a solvent on morphology and size of prepared particles is presented.

EXPERIMENTAL

All reagents were analytical grade. To avoid hydrolysis on storage, fresh stock solutions were prepared from $Zn(NO_3)_2 \times 6H_2O$ (Aldrich) and urea (Aldrich) in water. The experiments were carried out under the steering in 250 ml open reactors. As solvent we used water and mixture of water (W) and ethylene glycol (EG), diethylene glycol (DEG), tetraethylene glycol (TEG), polyethylene glycol (PEG, $M_w = 400$) and 1-propanol (1-P) respectively. Constant volume ratio (water/ polyols was 1/1) except for water/EG, where we had 1/3 and 3/1 ratio. During these experiments, the temperature was measured with Fluke 54 II thermocouple type K with thickness of 1 mm.

In all experiments the concentration of urea was kept constant 0.05 mol dm^{-3} and the concentration of Zn^{2+} ions was 0.01 mol dm^{-3}. The resulting solids were characterized by scanning field emission electron microscopy (SEM, Zeiss Supra 35 VP with EDS analyzer), while X-ray diffraction analyses (XRD) were carried out on a Siemens D-500 X-ray diffractometer. The origin of additional shoulder on XRD (Fig. 4) W:EG was from $K\alpha_2$ (XRD spectrum was not filtered). IR spectra were obtained on FTIR spectrometer (Perckin Elmer 2000) in the spectral range between 4000 and 400 cm^{-1} with spectral resolution of 2 cm^{-1} in transmittance mode. KBr pellets technique was used for sample the preparation. Particle sizes and distributions were determined from SEM images using CorelDRAW software. At least 100 crystals were measured per sample.

RESULTS AND DISCUSSION

Figure 1. SEM micrographs of samples after 4 hours of synthesis in different solvent: a - water, b - W/EG, c – W/DEG, d – W/TEG, e – W/PEG, f – W/1-P. Volume ration of W/polyol is 1/1 in all experiments.

In Figure 1 the SEM micrographs of six typical samples prepared after four hours of synthesis in different media water, water/EG, water/DEG, water /TEG, water/PEG and water/1-P are presented. While the Fig. 1a shows the formation of rod-like particles and some smaller flake-like particles, much more uniform formation of rods can be observed in b, c, and d of

Figure 1, respectively. All these samples were synthesized at the same reaction temperature (90°C), same concentration of Zn^{2+} and urea. These structures were reproducible when the synthesis was performed under the same conditions. In pure water, the mixture of water/EG and water/DEG (volume ratio was 1:1) longer hexagonal rods were obtained with smaller diameter. When instead, the mixture of water/TEG and water/PEG was used, rods much shorter and wider in diameter were synthesized.

The mechanism of the formation of these particles is debatable. In the paper Hu et al. (14) ZnO nanoparticles were prepared from Zn(II) acetate and NaOH in a series of n-alkanols (from ethanol to 1-hexanol). Their results show that nucleation and growth in ethanol and 1-propanol were retarded, compared to longer chain length alcohols where nucleation and growth were faster. Also the coarsening rate constant increased with temperature at longer solvent chain length, which was explained with the influence of solvent viscosity, surface energy and bulk solubility of ZnO. Due to all presented results, they concluded that the solvent presents an important parameter in controlling the particle size.

An XRD spectrum of our sample, prepared in pure water (Fig. 1a) is presented in Fig. 2. XRD analysis confirms the presence of pure ZnO (JCPDS 00-036-1451) and some $Zn_5(OH)_6(CO_3)_2$ (JCPDS 00-019-1458).

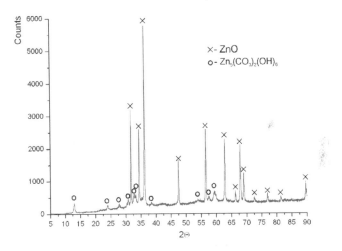

Figure 2. XRD pattern of sample prepared in water after four hours of synthesis.

FTIR spectra in Fig. 3 of the same sample confirm the presence of carbonate groups. The splitting of v_3 frequency of bands at 1512 and 1387 cm^{-1} were used to confirm microstructure changes in $Zn_5(OH)_6(CO_3)_2$ for samples prepared in autoclaves from $Zn(CH_3COO)_2$ with $(CH_3)_4NOH$ (15). They obtained that with longer heating time up to 72 hours at 160°C, much more crystalline $Zn_5(OH)_6(CO_3)_2$ was obtained that shows changes, visible in IR spectra as additional separation and sharpening of spectral line. As a typical example v_3 frequency of carbonate group is split into three pares at 1511 and 1387 cm^{-1}, 1549 and 1362 cm^{-1} and 1592 and 1337 cm^{-1}. Our sample (Fig. 3) shows only bands at 1517 and 1385 cm^{-1} and 1517 and 1422 cm^{-1}

which confirms that carbonate is not a good crystalline and that it is present also in amorphous state. The same observation can be confirmed from XRD spectra, where the intensities of $Zn_5(OH)_6(CO_3)_2$ are much lower than the intensities obtained from ZnO. In the IR spectrum (Fig. 3) the bands at 1045, 832 and 709 cm^{-1} additionally confirm the presence of carbonate groups that correspond to v_1 frequency. In literature, is presented that IR spectra of ZnO which show characteristically bands in the region from 680 up to 300 cm^{-1} were used not only for the characterization but also for the conformation of the shape of ZnO particles (16). In the work of A. Verges et al. (17), it was shown that the IR band for spherical particles was obtained at 458 cm^{-1} and it splits in two bands when the shape was changed. For prismatic microstructure they obtained two well split bands at 512 and 406 cm^{-1}. IR spectra of our sample presented in Fig. 3 show two bands at 509 and 466 cm^{-1} that confirm the presence of rod-like particles, as shown in Fig. 1a.

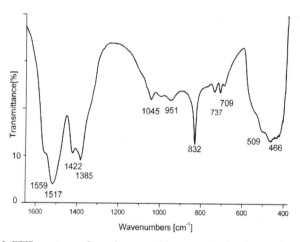

Figure 3. FTIR spectrum of sample prepared in water after four hours of synthesis.

XRD spectra of samples prepared in water/EG, water/DEG, water/TEG and water/1-propanol only confirm the formation of pure ZnO (Fig.4). IR spectra (not presented) of the same samples show only bands in the range from 700 up to 300 cm^{-1} that confirms formation of ZnO. The characteristic bands for ZnO are obtained at 564 and 472 cm^{-1}, 568 and 395 cm^{-1}, 519 cm^{-1} and 528 and 390 cm^{-1} for the mixture of water/EG, water/DEG, water/TEG, and water/1-P respectively. The shifting of the characteristic bands is connected with the shape of prepared particles (17).

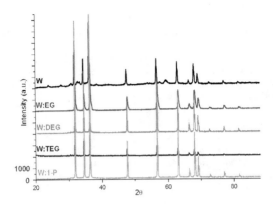

Figure 4. XRD patterns of samples prepared in different media after four hours of synthesis

The morphologies of all our prepared samples as shown in Fig. 1a-f consist predominantly of ZnO bipods. Each bipod consists of two rods that are linked in line that presents some novel morphology of ZnO (that is different from ZnO nanorods and nanowires (18) obtained from $Zn(NO_3)_2$ with hexamethylenetetramine (HMT). The average diameter of these bipods changes in dependence of polyols that were used as reaction media. The longer bipods were obtained in water/EG = 1:1 with the average length of 8 μm and the average diameter of 700 nm. When the mixtures of water with DEG, TEG and PEG were used, the size changed. The average length was 8, 4 and 1 μm in the mixture of water with DEG, TEG and PEG, respectively. The average diameter of rods/bipods increased, and in TEG, the size of diameter was almost as the length. The junction of obtained bipods was much more visible for the particles that grow in the water/TEG. In a few cases, as presented in Fig. 1c, some multipods like tetrapods were observed.

Hu et al. (18) presented a new way of formation linked ZnO rods (bipods, tripods, tetrapods) which was performed by ultrasonic irradiation and microwave heating (18) from zinc nitrate and HMT. The formation mechanism of ZnO rods linked as single crystalline bipods was proposed. The oriented attachment mechanism was suggested for the linkage of ZnO rods. That mechanism involved crystal growth by crystallographically controlled addition of primary individual crystals. They showed that formation of single crystalline bipods cannot be explained in their case by the Ostwald ripening process, where the large particles grow on the expense of smaller particles.

ZnO is a polar crystal and it crystallizes in Wurtzite structure with space group $P6_3mc$ and can be described as hexagonal close packing of O and Zn atoms with Zn atoms in tetrahedral sites. In Wurtzite structure there is no center of inversion (19). An inherent asymmetry along the c axis is present that allows anisotropic growth of the crystal in the 0001 (c-axis of crystal lattice) directions. The shape of formed nanorod with length and diameter (aspect ratio) of crystals are determined by the relative rates of growth of its various faces (20). The growth rate, in general, is controlled with some external factors like temperature, supersaturation, the influence of solvent, and impurities. The internal, structurally-related factors, as dislocation and intermolecular bonding controll the rate of growth. In our case, the external factors, the effect of solvent on nucleation and growth of crystals is studied.

As presented in Fig. 1a-f, we infer that different aspect ratios from 20 up to 1 of ZnO rods result from different growth rates in different reaction media. In our case, the growth rate along c-axis in water/EG is higher than in water/TEG. The detailed theoretical simulation of interface solvent interaction as suggested by Samulski et al. (20) with using a known parameter like surface energies of different crystal faces and solvent properties can verify the growth mechanism. The morphology of ZnO was controlled in our experiments with the chosen solvent. The particle growth and coarsening are strongly reliant on the chosen solvent through their viscosity and bulk solubility (14). The solubility of ZnO in dependence on the dielectric constant of the solvent was determined for ZnO in water/2-propanol mixture. The solubility of the ZnO particles increased with increasing dielectric constant which indicates that the partially covalent nature of bonding in ZnO plays an important role. The measured Zn(II) concentration was 49 μM and when water concentration was less than 60 vol%, the Zn(II) concentration was below the detection limit. A similar observation can be applied for our experiments, since the samples of ZnO prepared in pure water started to dissolve in some cases.

To understand the formation of ZnO, one must be acquainted with the aqueous chemistry of zinc and its hydrolysis. When $Zn(NO_3)_2$ is dissolved in water, Zn^{2+} cations, are octahedrally coordinated and form hexaqua ion $Zn(OH_2)_6^{2+}$. The ionic radius of Zn^{2+} is 88 pm and polarizing power (the ability to withdraw electron density) is weak. Due to this fact, as it was reported (21), at low pH (1-5), which was in our experiments (pH was around 5), hydrated zinc is stable. From partial charge model (PCM) (22) it can be estimated charge and molecular species of Zn^{2+} at defined pH. Zinc species, i.e. $Zn(OH_2)_6^{2+}$, destabilize with increasing pH and undergo hydrolysis. With increasing temperature, urea decomposed and OH concentration increased. PCM predicted that at pH around 7 zinc-coordinated water molecules will deprotonate and form $Zn(OH_2)_4(OH)_2$. Further on, through condensation reactions this species can eliminate water that leads the formation of metal oxide. The solubility is reduced and nucleation occurs.

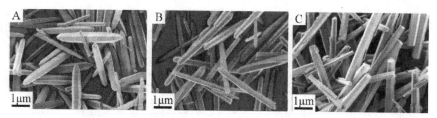

Figure 5. SEM micrographs of samples prepared in water after different time of synthesis: a- 30 minutes, b- 45 minutes, c- 60 minutes.

ZnO crystal is a polar crystal with positive polar plane rich in Zn and the negative polar plane rich in O (24). In general in the hydrothermal process, where precipitation was performed with strong base (23-25), it was reported that $[Zn(OH)_4]^{2-}$ presented the growth unit of ZnO that leads to the different growth rate of planes, i.e. $V_{(0001)} > V_{(01\bar{1}\bar{1})} > V_{(01\bar{1}0)} > V_{(01\bar{1}1)} > V_{(000\bar{1})}$ (25). It is known that if the growth rate is quicker the disappearance of the plane is more rapid. Due to that 0001 plane disappears in the hydrothermal process and pointed shape at the end of existed c axis.

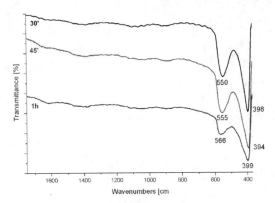

Figure 6. FTIR spectra of sample prepared in water after 30, 45 and 60 minutes of synthesis

The formation of these types of bipods was observed after 30 minutes of reaction when the solvent was pure water (Fig.5a). With further heating (45 min), as presented in Fig. 5b, the pointed shape at the end of bipods lost the sharpness, and after one hour, the plain shape (Fig. 5c) was observed. IR spectra of these samples (Fig.6) show two well resolved bands of all prepared samples that are characteristic for rod-like particles (in the range from 600 up to 400 cm^{-1}). With time, for example after four hours, as presented in Fig. 1a, two types of particles were observed, big rod-like and smaller flake-like particles were formed that can explain that due to the high dielectric constant of water, ZnO starts to dissolve, forming zinc hydroxyl carbonate.

With substituting part of water (in volume ratio) with EG, nice rod-like particles of ZnO were formed after 30 minutes (Fig.7). It is clearly visible that the size of rod like particles is closely connected with volume ratio of water/EG. Bigger particles were formed in pure water after 30 minutes (as presented in Fig.5a). The size of particles decreased with increased volume /concentration of added EG (Fig.7). The smallest particles were formed after 30 minutes at volume ratio of water/EG=1/3 (Fig.7g). After four hours (Fig. 1b) at equal volume ratio of water/ EG these rods' average size is 4.4 μm in length and 330 nm in diameter. At water/EG =3/1, the inhomogeneous formation of nano-rods with different size is shown on Fig. 7c (the average length is 4 μm and diameter 400 nm). With the opposite volume ratio i.e., three parts of EG and one part of the water obtained rods are shorter (less than 2 μm) and much more narrow, with diameter of about 250 nm (Fig.7i) It means that with changing the ratio of polyol and water in our case EG the size of rod-like particles can be controlled. A similar observation was obtained (24) for controllable synthesis of polyoxometalate (POM) nanocrystals in PEG/H$_2$O media. The diameter of prepared nanorods increased with increasing water content. In the liquid systems with addition of water the linear reaction field presents reaction space that become larger, and due to that, nanorods will have a larger diameter.

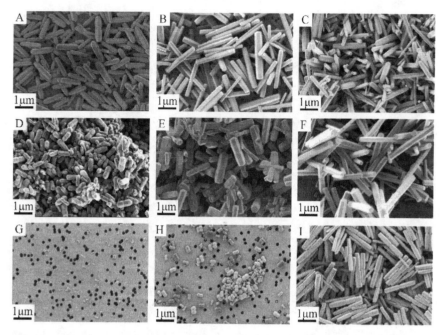

Figure 7. SEM micrograph of samples prepared in W/EG = 3/1; a - 30 minutes, b - 45 minutes, c- 240 minutes, W/EG = 1/1; d - 30 minutes, e - 45 minutes, f - 240 minutes and W/EG: = 1/3 g - 30 minutes, h - 45 minutes and i- 240 minutes.

CONCLUSION

One step low temperature solution phase preparation method was successfully used for the preparation of submicrometer-sized ZnO. The main results of our work show that the growth of our particles can be controlled with changing the type of polyol that serves as a media for the preparation of ZnO. With substituting a part of water with EG, DEG, TEG, PEG and 1-P the size of rod-like particles (length and diameter) of ZnO was changed. The size of rod-like particles is also connected with volume ratio of water/EG. Bigger particles were formed in pure water and the size of particles decreased with increased volume/concentration of added EG. The smallest particles were formed after 30 minutes at volume ratio of water/EG=1/3. At water/EG =3/1 the average length is 4 μm and diameter 400 nm. With opposite volume ratio i.e., three parts of EG and one part of the water obtained rods are shorter (less than 2 μm) and much more narrow, with diameter of about 250 nm. It means that with changing the ratio of EG and water the size of rod-like particles can be controlled.

ACKNOWLEDGMENT
The authors gratefully acknowledge the financial support of the Ministry of Higher Education, Science and Technology of the Republic of Slovenia, and the Slovenian Research Agency (program P1-0030, project J2-6027).

FOOTNOTES
*Corresponding author zorica.crnjak.orel@ki.si

REFERENCES
1   H. Yu, Z. Zhang, M. Han, X. Hao and F. Zhu, J. Am. Chem. Soc., **127**, 2378-2379 (2005)
2   (a) M.H. Huang, S. Mao, H. Feick, H. Yan, Y. Wu, H. Kind, E. Weber. R. Russo, P. Yang, Science **292**, 1897-1899 (2001)
3   D. Ledwith, S.C. Pillai, G.W. Watson, J.M. Kelly, Chem. Comm., 2294-2295 (2004)
4   J. Zhang, L. Sun, C. Liao, C. Yan, Chem. Comm., 262-263 (2002)
5   Z.R. Tian, J.A. Voigt, J. Liu, B. Mckenzie, M.J. Modermott, J. Am. Chem. Soc. **124**, 12954-12955 (2002)
6   Y. Zhang, H. Jia, X. Luo, X. Chen, D. Yu, R. Wang, J. Phys. Chem. B, **107**, 8289-8293 (2003)
7   X.Y. Kong, Z.L. Wang, Nano Lett., **3**, 1625-1631 (2003)
8   C. Wang, E. Shen, E. Wang, L. Gao, Z. Kang, C. Tian, Y. Lan, C. Zhang, Materials Letters, **59**, 2867-2871 (2005)
9   L. Poul, S. Ammar, N. Jouini, E. Fievet and F. Villain, Journal of Sol-Gel Sci. and Technol., **26**, 261-265 (2003)
10 Z. Crnjak Orel, E. Matijević, D.V. Goia, J. Mater. Res., **18**, 1017-1022 (2003)
11 A. Anžlovar, Z. Crnjak Orel, M. Žigon, J. European Ceramic Society, **27**, 987-991 (2007)
12 Z. Crnjak Orel, A. Anžlovar, G. Dražić, and M. Žigon, Crystal Growth & Design 7, 453-458 (2007)
13 M. Castellano and E. Matijević, Chemistry of Materials, **1**, 78-82 (1989).
14 Z. Hu, G. Oskam and P. C. Searson, J. Colloid and Inteface Sci. **263**, 454-460 (2003).
15 S. Musić, S. Popović, M. Maljković, Đ. Dragčević, J. Alloys and Compounds **347**, 324-332 (2002).
16 S. Hayashi, N. Nakamori and H. Kanamori, J. Phys.Soc. Japan **46**, 176-183.
17 M.Andres Verges, A. Mifsud and C.J. Serna, J. Chem. Soc. Faraday TRans., **86**, 959-963 (1990)
18 X.-L. Hu, Y.-J. Zhu, S.-W. Wang, Materials Chem. and Phys., **88**, 421-426, (2004)
19 L. Vayssieres, K. Keis, S.-E. Lindquist, and A. Hagfeldt, J. Phys. Chem. B 105 (2001) 3350-3352.
20 B. Cheng and E.T. Samulski, Chem. Comm., 986-987 (2004)
21 D. Kisailus, B. Schwenzer, J. Gomm, J.C. Weaver and D.E. Morse, J.A.C.S., **128**, 10276-10280 (2006)
22 M. Henry, J.P. Jolivet, J. Livage in: Chemistry, Spectroscopy and Applications of Sol-Gel Glasses; R. Reisfeld, Springer-Verlag, New York, Vol 77, p. 153. 1992
23 H. Zhang, D. Yang, Y. Ji, X. Ma, J. Xu and D. Que, J. Phys. Chem. B, **108**, 3955-3958 (2004)
24 Z. Kang, E. Wang, M. Jiang, S. Lian, Y. Li and C. Hu, Eur. J. Inorg. Chem. 370-376 (2003)
25 W.-J. Li, E.-W. Shi, W.-Z. Zhong and Z.-W. Yin, J. Crystal Growth, **203**, 186-196 (1999)

POROUS AND DENSE PEROVSKITE FILMS

A. Pohl*[1], G. Westin[1], and M. Ottosson[1]
[1]Department of Materials Chemistry, Ångström Laboratory,
Uppsala University,
SE-751 21 Uppsala, Sweden
pohl@mkem.uu.se

A. Grishin[2], S. Khartsev[2], and R. Fors[2]
[2]Condensed Matter Physics, Royal Institute of Technology,
SE-164 40 Stockholm-Kista, Sweden

ABSTRACT

$La_{0.5}Sr_{0.5}CoO_3$ (LSCO) and $La_{0.67}Ca_{0.33}MnO_3$ (LCMO) films were prepared by sol-gel techniques and their structural and transport properties investigated. Films were spin-coated onto (001) $LaAlO_3$ (LAO), (001) $SrTiO_3$ (STO), $Pt/TiO_2/SiO_2/Si$, and Al_2O_3 substrates, and heated to 800°C. The structural properties were investigated using X-ray diffraction, scanning and transmission electron microscopy (SEM and TEM). SEM studies showed that the films were crack-free and adhered well to the substrates. No preferential orientation of the perovskite films was observed on Si- or $Pt/TiO_2/SiO_2/Si$-substrates, but films deposited on LAO and STO showed good alignment with the substrate. Transport measurements of epitaxial LCMO films show maximum temperature coefficient of resistivity (TCR) of 6.1 % K^{-1} at 241 K and colossal magnetoresistance (CMR) of 32 % at 246 K. The conductivity LSCO polycrystalline film was 1.7 mΩcm, while a epitaxial film had a conductivity of around 1.9 mΩcm.

INTRODUCTION

Mixed valence perovskite exhibit a wide verity of interesting chemical, physical, and structural properties, which has made them a dynamic research field for many years. As an example, manganites like $La_{0.67}Ca_{0.33}MnO_{3+x}$ (LCMO) are very interesting both from a fundamental physics standpoint and due to their promise for potential application in various devices such as un-cooled infrared (IR) bolometers and field effect transistors (FET) [1-6]. Spin-dependent transport close to the para-to-ferromagnetic transition (semiconductor-to-metallic) temperature, T_c, causes the resistivity to strongly depend on magnetic field (colossal magnetoresistance, CMR) and temperature.

$La_{1-x}Sr_xCoO_3$ (LSCO) mixed-valence perovskites may have smaller CMR than the LCMO, but it show several interesting properties such as good electronic and ionic conductivity, and catalytic properties.[7-10] The composition with x = 0.5 has a slight rhombohedral distortion of the perovskite structure and a low resistivity at room temperature, with reports of down to ca 100-300 μΩcm for the bulk materials [11-16] and films [16-19]. It is therefore of great interest as a ceramic electrode or buffer layer for perovskite-based memories, sensors and actuators with good bonding properties and low fatigue [20] and hence plentiful studies have been made on preparation of LSCO films.[16–19,21–26]

To date, most perovskites engineered towards applications are grown by high-vacuum physical techniques which is not an industrially viable technique since it requires relatively sophisticated and expensive equipment and there is limited possibility of large area deposition. Sol-gel derived thin films do not suffer these drawbacks and represent a cost effective and versatile route to complex-composition thin films. The outcome of sol–gel and related solution-based processing routes, such as chemical solution decomposition/metal organic decomposition (CSD/MOD), is highly dependent on the choice of precursors and on the heat-treatment program to remove residuals and to achieve crystallisation of the target phase. Therefore a careful study including all steps from the precursors to the phase development during the conversion of the gel to the oxide is essential for the evaluation and exploitation of a solution-based process. But in spite of this, rather few studies have been made in this area. We have previosly described alkoxide-based routes to $La_{0.5}Sr_{0.5}CoO_3$ and $La_{0.67}Ca_{0.33}MnO_3$ nano-phase powders, using methoxy-ethoxides of Mn, Co, La, Ca and Sr [27,28]. The chemistry and phase development during conversion of the homogeneous oxo-carbonate gels obtained by air hydrolysis to the final nano-phase perovskite was described in detail, and it was found that temperatures down to 700°C, could be used without annealing.

It is however of interest to further study both polycrystalline and epitaxial high-quality films as substrates for various kinds of hetero-layered perovskite films, and hence this study focuses on the formation of LCMO and LSCO films using the alkoxide sol-gel routes.

EXPERIMENTAL SECTION
Synthesis

All synthesis and handling of precursors were made in a glove box, with an atmosphere of dry, oxygen-free argon. The solvents were distilled over CaH_2 under inert atmosphere. The lanthanum, calcium, and strontium precursors were prepared by direct reaction of the metal with methoxy-ethoxide (moeH). In the case of La, ca 0.5 mg of $HgCl_2$ was used as catalyst. The resulting alkoxide solutions contained suspended fine particles, which were removed by centrifugation. The manganese and cobalt precursors were prepared by metathesis of anhydrous metal chlorides and potassium methoxy-ethoxide at ambient temperature. After 24 hours of stirring, the mixture was centrifuged to separate the solid KCl from the alkoxide solutions. The obtained raw-product frequently contained small amounts of K and/or Cl, according to SEM–EDS analysis, and therefore the alkoxides was purified by recrystallisation. The precursors were mixed in solution in stoichiometrical ratios with respect to the target perovskites; $La_{0.67}Ca_{0.33}MnO_3$ (LCMO), and $La_{0.5}Sr_{0.5}CoO_3$ (LSCO).

For the studies on the phase development during heat-treatment, gel powders were prepared by depositing mixed alkoxide solutions onto aluminum foil in air. After 30 minutes the gels were dry and could easily be removed from the foil. Heat treatment to study the gel to oxide transformation process was performed with a TG apparatus, samples were heated in air to different temperatures and analyzed by PXRD, FT–IR spectroscopy and TEM-EDS. A detailed account of the phase developments are provided elsewhere [27,28], here only the aspects of the processes that are of importance in connection with the film studies are discussed.

Films were spin-coated onto Si, $Si/SiO_2/TiO_2/Pt$, polycrystalline α-Al_2O_3, and $LaAlO_3$ (LAO) or $SrTiO_3$ (STO). Solutions with total metal concentrations of 0.2–0.6M were used, and spinning rates of 3000–3500 rpm, 30s. The gel films were converted to oxide by heating in air in a tube furnace. Based on the phase development studies, 800°C was chosen as final temperature for the heat-treatment. For LCM heating rates of 2-100°C·min^{-1} were used, for LSCO 2-5°C·min^{-1}. Some films were post-annealed at 800 or 1000°C. The film on α-Al_2O_3 was also heated to 1000°C. Before conductivity measurements were made, the LSCO films were annealed at 750°C for 15 min, under an oxygen atmosphere in accordance with the literature [29-31].

Equipment

A transmission electron microscope equipped with an energy dispersive X-ray spectrometer (TEM–EDS; Jeol 2000FX–Link 10000AN) was used for studies of the habit, metal ion homogeneity and crystallinity of films scraped off the substrates. A field emission scanning electron microscope (SEM; LEO 1550) was used for studies of the microstructure of the films. The polycrystalline films on α-Al_2O_3 were examined by grazing incidence X-ray diffraction (GI-XRD) measurements with a Siemens D5000 diffractometer, using Cu Kα radiation, while the epitaxial films on STO were investigated with ω-2θ scans, rocking curves, and reciprocal space mapping (RSM) using a Philips MRD with a parallel beam set-up, with x-ray mirror (Ni/C) as primary and secondary optic. Transport properties were measured by a standard 4-point probe technique using a Jandel apparatus. In the case of LCMO the measurements were performed in an electromagnet, capable of fields up to 0.7 T.[32] By switching the current $+-+$, effects of thermoelectric voltages were eliminated. The temperature coefficient of resistivity is defined as $TCR \equiv d \ln \rho / dT$ and the magnetoresistance ratio as $MR \equiv (\rho_0 - \rho_{0.7T})/\rho_0$.

RESULTS AND DISCUSSION

Solution processing

Precursor system While there are several soluble La and Sr alkoxides, most simple cobalt and manganese alkoxides form insoluble polymers [33-37]. In this study, the Mn and Co sources used was the methoxy-ethoxides, which is highly soluble in common organic solvents, such as moeH. There has been no structure reported for a Co-methoxy-ethoxide or oxo-methoxy-ethoxide, and we have not yet succeeded in preparing single crystals for structural analysis. The structure of the Co-methoxy-ethoxide might be similar to that of the corresponding disk-shaped Mn-methoxy-ethoxide $[Mn_{19}O_{12}(moe)_{14}(moeH)_{10}]$ (Mn19), which has an MnO-like $Mn_{19}O_{12}$-core, surrounded by organic groups that make it soluble [38]. The structure is shown in figure 1. The oxo oxygens of Mn19 were obtained by auto-decomposition of alkoxo ligands, and this can be expected to be facile also for the Co-moe compound.

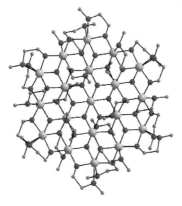

Figure 1. The molecular structure of $[Mn_{19}O_{12}(moe)_{14}(moeH)_{10}]$.

The advantage of alkoxides compared to *e.g.* acetyl-acetonates, acetates or citric acid complexes, is that alkoxides are far more reactive and form stable oxygen bonds between the constituents already in the gel. The high reactivity also results in a very low amount of residual organic groups since most alkoxo-groups are removed in the hydrolysis-condensation reactions and the alcohols formed can be evaporated. Furthermore the lower reactivity of acetate precursors can lead to phase separation within the gels of mixed acetate-alkoxide systems.[36] Thus, alkoxide routes yield more homogeneous and pure gels, allowing for well-controlled low temperature conversion of the gel to oxide.

The simplicity of the alkoxide system also makes it fast and easy to prepare films or powders of different compositions. After determining the concentrations of the different alkoxide solutions, they are mixed stoichiometrically to the desired composition(s). The mixed solution is then ready to be used; there is no need for sol preparation, addition of a gelling agent or heating of reaction mixture, as in many other sol–gel systems.[39] The high reactivity of the alkoxides to the moisture in the air enables formation of very pure gel films when spin-coated.

The precursor synthesis, especially of the La, Ca, and Sr precursors, is also much more simple and straight forward than e.g. in the proponic-acid route, which involves distillation and drying/redissolution steps, and also requires filtering of the mixed precursor solution before it can be used.[40] This also reduces the compositional control of the proponic-acid route compared to the alkoxide route.

Preparation temperature We have seen that our all-alkoxide sol–gel route yielded hydrated oxo-carbonate gels with no or only small amounts of organic groups remaining, and that the gels were homogeneous in their elemental composition. Heating in air yielded the phase pure perovskite at comparatively low temperatures, even without annealing; *i.e.* 690°C (LCMO) and 700°C (LSCO) for heating-rates of $2°C \cdot min^{-1}$.

The fact that no annealing time is needed could make it possible to obtain low-temperature modifications of the perovskite. This was seen in the LCM study where the perovskite samples where divided between two types of orthorhombic modifications, depending on the lattice parameters, which differed mainly in the b_0-axis length. The shorter b_0-axis of the samples heated to lower temperatures is due to a high oxygen content and a lower amount of larger and Jahn-Teller distorted Mn^{3+} ions. The larger b_0 lattice parameter for samples heated to higher temperatures is similar to data reported for compositions of $La_{0.67}Ca_{0.33}MnO_{3.057}$ – $La_{0.67}Ca_{0.33}MnO_{3.004}$,[41] and by assuming the composition of our LCMO sample heated to 1000°C to be $La_{0.67}Ca_{0.33}MnO_{3.0}$ the oxygen stoichiometry at lower temperatures could be estimated from the TG graph. Thus a LCM sample heated to 800°C $(20°C \cdot min^{-1})$ would have the composition $La_{0.67}Ca_{0.33}MnO_{3.28}$. It should however be emphasized that this is a rough approximation, and that no exact determination of oxygen stoichiometry has been made for any sample in this study.

The reduction of oxygen stoichiometry on prolonged annealing or heating to higher temperatures, is a feature that distinguish the all alkoxide sol–gel route from e.g. PLD and solid-state synthesis, where post-annealing is known to increase the oxygen content. The perovskite initially formed by the all-alkoxide route has a higher oxidation state than found in many solid-state samples even after annealing in oxygen.

No large variations in cell-parameters on annealing were observed for LSCO, but the cubic cell becomes hexagonally distorted, and the TG-graph indicated some oxygen loss between 700 and 1000°.

The gel and the evolution on heating The origin of the initial high oxidation state of the LCMO perovskite could at least partly be attributed to the reactive manganese alkoxide precursor. When the mixed precursor solutions were subjected to air, the oxidation of the Mn^{2+} was immediate and caused a darkening of the solution. Taking into account the different reactions occurring on heating as indicated by IR spectroscopy, the assumption that oxygen stoichiometry at 1000°C would be $La_{0.67}Ca_{0.33}MnO_{3.0}$ allowed an approximate composition of the gel to be estimated from the TG graph. The obtained molecular formulas indicate the average oxidation state of the manganese to be +4. For LSC-system oxidation of the divalent Co-ions of the the alkoxide precursor also occurred during gelation in air; the spectroscopic studies of the LSC-gel indicted a mixture of di- and trivalent Co ions.

The LCM-gel was completely hyrdolysed, but the LSC-gel occasionally contained small amounts of organic residue, but roughly the gels can be described as amorphous hydrated oxo-carbonates. The carbonate groups stem from absorption of CO_2 from the air, and the decomposition of the carbonates is the limiting step for the perovskite formation. The final carbonate decomposition occurred between 600 and 750°C, depending on heating-rate and composition. Annealing at temperatures lower than this, can if long enough decompose the carbonates almost completely. Thus isothermal heat-treatment can yield perovskite at even lower temperatures than 600°C. However, this leads to A-site inhomogeneity or deficiency of certain ions in the perovskite, probably due to that carbonate decomposition releases different ions at different rates.

Films

Based on the phase development studies, 800°C was chosen as standard preparation temperature for the heat-treatment of films. Polycrystalline films with random crystal orientation were obtained on Si, Si/SiO_2/TiO_2/Pt, and polycrystalline α-Al_2O substrates[3], while LCMO films on $LaAlO_3$ (LAO) and LSCO films on $SrTiO_3$ (STO) showed good alignment with the substrate.

The thickness of the films could be varied (Fig 2) through the concentration of the precursor solution, but multideposition was required to obtain thick films without cracks. For both LCMO and LSCO, single deposition resulted in final film thicknesses of 20–30 nm and 80–100 nm with the 0.2 and 0.6M solution, respectively. For LCMO a *ca* 250 nm thick film were prepared from a 0.6 M solution by three depositions with intermediate heating.

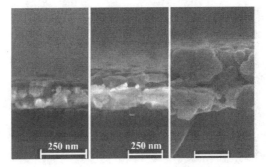

Figure 2. Cross-section SEM images of LCMO film on Pt/TiO_2/SiO_2/Si substrate made by single depositions of a 0.2 M solution (left) and a 0.6 M solution (middle), and by three depositions with intermediate heating to 800°C of a 0.6 M solution (right).

Effect of heat-treatment on porosity The alkoxide precursors used yielded fully hydrolyzed gel films, often without any organic parts left, thus reducing the gas formation during heating and thereby the risk of cracking. However the gel contain carbonates (from absorption of atmospheric CO_2), and when the carbonates decompose there is a evolution of CO_2 which is expected to cause some porosity, and if too fast perhaps also cracks. The effect of heating rate was evaluated in the LCM study, where heating rates up to $100°C \cdot min^{-1}$ were used for films deposited on Pt substrates. It was found that heating rates between 2 and $20°C \cdot min^{-1}$ resulted in smooth crack-free films, whereas some large pores and bubbles were observed in the film heated at $50°C \cdot min^{-1}$, and the film heated at $100°C \cdot min^{-1}$ was completely covered by 1-2μm sized bubbles.

The films heated to 800°C were usually porous, however there is a clear tendency for films heated at lower rates to be less porous than films heated at higher rates, and for $2°C \cdot min^{-1}$ quite dense films were obtained on $Pt/TiO_2/SiO_2/Si$ and Al_2O_3. Annealing might reduce the porosity, on a favorable substrate. The LCMO film on LAO-substrate initially heated to 800°C showed a dramatic change of the morphology after annealing for 2h at 1000°C, from a highly porous structure of *ca* 50-100 nm sized crystals to a dense film without visible grain boundaries (Fig 3). A different result of annealing a LCMO films was found for films deposited on $Pt/TiO_2/SiO_2/Si$-substrate and heated at $20°C \cdot min^{-1}$. In this case annealing caused growth of larger pores at the expense of smaller, resulting in a poor coverage of the surface (Fig 4). This might be en effect of the Pt-structure, but further studies are needed to verify this.

LSCO films on (100) STO substrates annealed for 30 min at 800°C, had a morphology somewhat intermediate the two LCMO/LAO films; the surface was typically dense and smooth but square shaped crystalline features in the order of 30 nm could still be distinguished (Fig 3). Some small holes in the film, shaped after the epitaxial growth directions could be seen, as well as some areas with a more porous epitaxial structure in minor parts of the films.

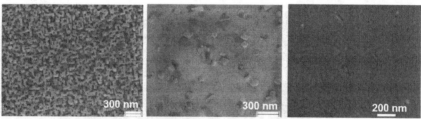

Figure 3. LCMO film on LAO substrate heated to 800°C (left), the same film after annealing at 1000°C, 2h (middle), and the LSCO film on STO annealed at 800°C, 30 minutes (right).

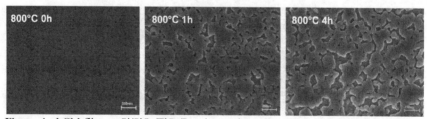

Figure 4. LCM film on $Si/SiO_2/TiO_2/Pt$ substrate heated to 800°C (left) and the same film after annealing in air at 800°C for 1h (middle) and 4h (right). Scalebar 250 nm.

20–150 nm sized round pinholes were found in the LSCO film deposited on STO and Al_2O_3. These could be induced by the substrate, or it may be that water and carbon dioxide gases that are formed during the xero-gel decomposition blow holes in the viscous film before the final formation of LSCO. The frequent observation of substrate grain boundaries or triple junctions in the pinholes for the LSCO/Al_2O_3 film can then be explained by gas more easily collecting in these depressions and blowing holes in the film. By further optimisation of the spin-coating, heating cycle and/or chemical activation of the substrate surface it should be possible to reduce these defects to a very low level.

Crystallites on top of the smooth film surface were observed for the annealed LSCO/STO and LCMO/LAO films. The mechanism by which they form is not clearly understood at this point, but this phenomenon is only observed for the epitaxial films. This kind of surface grown crystals are also observed in perovskite films obtained by other techniques, such as high-vacuum physical techniques.

Crystal structure and transport properties The GI-XRD scan of LSCO films on Pt/TiO_2/SiO_2/Si-, and α-Al_2O_3 substrates showed only peaks that could be associated with the LSCO perovskite phase, and that the films were polycrystalline with no obvious preferred orientation of the crystallites. The cubic cell dimensions were a_c = 3.825 Å for the film on α-Al_2O_3, and a_c = 3.84 Å for the film on Pt/TiO_2/SiO_2/Si. For LSCO with a $Sr_{0.5}La_{0.5}CoO_{3-\delta}$ composition, a slight rhombohedral distortion is expected [11,12], and a better fit of the diffraction data for the LSCO/α-Al_2O_3 film could be achieved with a hexagonal cell with a_h = 5.400(2) c_h = 13.283(7). No significant change in cell volume was seen after heating the LSCO/α-Al_2O_3 film to 1000°C, and the obtained hexagonal cell, a_h = 5.426(2) c_h = 13.254(4), can be compared to the a_h = 5.420(2) c_h = 13.315(8) obtained for a nano-phase powder prepared at 1000°C from the same precursors [27].

The θ–2θ XRD scans of the LSCO/STO films (Fig 5) revealed an epitaxial 001 cube-on-cube orientation of the LSCO film on the STO substrate. A small peak of the (110) orientation at 2θ = 33°, which is normally the strongest peak in polycrystalline LSCO, could be discerned from the noise. The in-plane a/b-axis was determined to be 3.894 Å for the 20 nm film and 3.897 Å for the 80 nm film. The out-of-plane c-axis was determined to be 3.789 Å for the thin film and 3.782 Å for the thick film, using the RSM of the asymmetrical reflection (024), shown in figure 6. This shows that the a-axis of the LSCO cell was stretched to almost perfectly match the larger STO cell having a cell dimension of 3.905 Å, while the c-axis was decreased to compensate for the a-axis lengthening. The 004 and 024 rocking curve full width at half maximum values were in the range 0.30–0.40, with somewhat broader peaks for the thicker film.

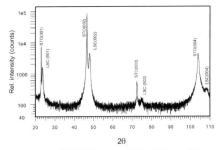

Figure 5. θ–2θ XRD scan of 80–100 nm film on STO.

The RSM pattern show that the thin film shows more broadening in the vertical direction compared to the horizontal direction (2θ broadening), which can be attributed to film thickness or inhomogeneous strain, or a combination of both. For the thicker film, an asymmetrical broadening was observed, indicating some relaxation for the thicker film.

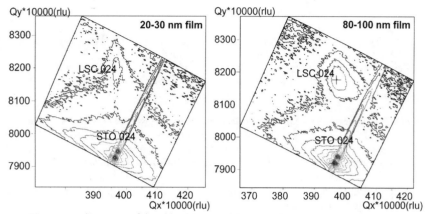

Figure 6. RSM scans of the; thin (20–30 nm) (a) and thick (80–100 nm) (b) films.

GI-XRD measurements were used to examine the phase purity of the LCMO films, and as expected from the phase development studies the perovskite was the only crystalline phase present. A more thoroughly XRD study of LCMO films on LAO substrates was performed in connection with the transport measurements, these films had a thickness of 50nm. The porous film heated to 800°C showed a preferred crystal orientation induced by the substrate. Calcination in oxygen at 1000°C for two hours not only resulted in a densification as observed by SEM, but also in improved epitaxial quality (Fig 7) and a decrease in the out-of-plane lattice parameter from 3.849(7) Å to 3.856(6) Å. The lattice parameter change is consistent with loss of oxygen, also observed in the phase-development studies on powder samples.

Transport measurements for the LCM/LAO films were performed at zero-field and at 0.7 T. Increasing peak TCR and MR with annealing could be expected due to increased domain sizes.[42] This was indeed observed, but unlike films deposited by pulsed laser deposition (for which post annealing leads to a decrease in resistivity and a peak shift to higher temperature due to incorporation of more oxygen) there was no significant peak shift in transport measurements for sol–gel film, and a slight increase in resistivity, at least below the peak position (Figure 8). This can at least partly be explained by the sol–gel film loosing oxygen during post annealing, but further investigations of transport properties, oxygen content and morphological evolution are needed to establish the origin of the behavior of the sol–gel derived films. Table 1 lists the transport properties sol-gel prepared LCMO/LSCO film, as well as a PLD film for comparison.

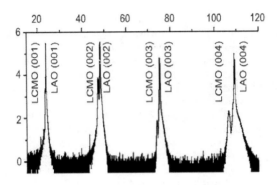

Figure 7. XRD 0-20 scan of post-annealed sol-gel LCMO on LAO, annealed in oxygen for 2h at 1000°C

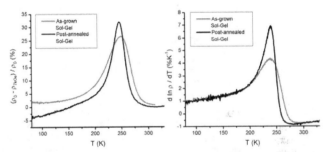

Figure 8. Transport measurements, MR (left) and TCR (right) for LCM films on LAO heated to 800°C with no annealing and post annealed in oxygen for 2h at 1000°C.

Table 1. Summary of transport properties.

Film	T_ρ^{peak} (K)	TCR^{peak} (%K^{-1})	T_{TCR}^{peak} (K)	MR^{peak} (%)	T_{MR}^{peak} (K)
As-grown PLD	268	4.6	249	27	249
Post-annealed PLD	274	8.2	258	35	263
800 °C sol-gel	269	4.4	239	26	249
Post-annealed sol-gel	258	6.1	241	32	246

The resistivity of the polycrystalline LSCO/Al$_2$O$_3$ films was of ca 1.6 mΩcm, which is in the same order as reported for other good polycrystalline films. The variation among different measurements was within 1%, indicating a homogeneous conductivity in different parts of the sample. The resistivity of the 20–30 and 80–100 nm thick epitaxial films were 30 mΩcm, and 1.9 mΩcm, respectively. The former varied within ca 3% and the latter varied within 1%.

Litterature reports on resistivities of epitaxial LSCO films shows great variations depending on processing method and conditions, and values as low as around 100–300 µΩcm have been given for LSCO films prepared by PLD on STO [16-19]. The higher resistivity in our epitaxial films, in spite of the seemingly good epitaxial structure, is not clearly understood, but it might have an origin in the high strain in the relatively thin films, which makes the cell deviate more in dimensions than is commonly reported in the literature. Thus the expanded a-axis, implying a poorer metal-oxygen overlap in the direction of conduction, could result in a lower conductivity. The slightly increased cell volume compared to the relaxed LSCO would also indicate a somewhat reduced average oxidation number compared to the unstressed cell, which should also be a reducing factor for the conductivity. However, a solution-derived epitaxial porous LSCO film on STO having only a slight tetragonal distortion, with $a = 3.82$ Å and $c = 3.84$ Å, has been reported to have a resistivity of 12.4 mΩcm [43], showing that micro-structural effects such as porosity also are of great importance. Other explanations can be found in the quite thin films, where surface effects may play a role, or the structural defects that will to some degree increase the conduction path length, since these thin films seem to form holes all the way down to the substrate. Thin films are known to be sensitive to surface reactions, but we haven not seen any important resistivity difference between a 1 year old sample and one recently heated, which indicates that this is not the major explanation of the reduced conductivity in our case. Therefore it seems likely that the expanded cell stemming from the good epitaxial matching with the larger STO cell is the main contributor to the increased resistivity in our LSCO films.

CONCLUSIONS

An all-alkoxide based solution route to high quality epitaxial LCMO and LSCO films on LAO and STO, respectively, has been described. On Pt/TiO$_2$/SiO$_2$/Si, and Al$_2$O$_3$ substrates the films became polycrystalline with no preferred orientation. For epitaxial films post-annealing resulted in decrease or removal of initial porosity, while annealing polycrystalline films led to increased pore- as well as grain-size.

The highly reactive and easily oxidized Mn-alkoxide precursor makes high oxygen-excess modifications of the LCMO available, and LCMO films on LAO prepared at 800°C showed an apparent oxygen excess, which was removed by post-annealing at 1000°C. High c-axis orientation together with strong in-plane texture indicates epitaxial quality of post-annealed films. Increasing peak TCR and magnetoresistance with annealing resulted from increased domain sizes, while slight increase in resistivity below the peak position and absence of significant peak shift might be due to the change in oxygen stoiciometry.

For 20–30 nm thick LSCO films, the cell was strained to closely adhere to the STO (001) surface, but compensated this in-plane cell dimension increase with a c-axis reduction, resulting in a cell volume close to that of the unstrained nano-phase powder. Thicker (80–100 nm) films were relaxed to some degree at the surface. The conductivity of the polycrystalline LSCO film on α-Al$_2$O$_3$ is comparable to other reported results, while the epitaxial films are, as far as we can find, better than the previous solution-derived films, but fall short by an order of magnitude of the best reported LSCO films prepared by physical high-vacuum techniques. In spite of this, the films are expected to be useful for assisting epitaxial perovskite growth on ceramic electrodes; the strained comparatively large LSCO cell should provide a good matching for many larger perovskites, and the resistivity is still quite low. Possibly, a lower resistivity can be obtained on smaller substrate lattices and with increased film thickness, but this may be less suitable for the matching of a larger perovskite cell on top of the LSCO film.

Thus the alkoxide sol–gel route can be used to prepare epitaxial LCMO films with transport properties comparable to those prepared by PLD, while epitaxial LSCO films makes up for lower conductivity by better possibilities for lattice matching, making it interesting for use as conductive substrates for electro-ceramic perovskites, as well as for other catalytic and ionic permeation materials of simple or complex shape.

Acknowledgement
This study was financed by the Swedish Research Council (VR).

REFERENCES

1. M. B. Salamon and M. Jaime, *Rev. Mod. Phys.* 73, (2001) 583.
2. W. Prellier, Ph. Lecoeur, and B. Mercey, *J. Phys.: Condens. Matter 13*, (2001) R915.
3. A. M. Haghiri-Gosnet and J. P. Renard, *J. Phys. D: Appl. Phys.* 36, (2003) R127.
4. M. Rajeswari, C. Chen, A. Goyal, C. Kwon, M. C. Robson, R. Ramesh, T. Venkatesan, S. Lakeou, *Appl. Phys. Lett.* 68, (1996) 3555.
5. A. Lisauskas, S. I. Khartsev and A. M. Grishin, *Appl. Phys. Lett.* 77, (2000) 756; 77, (2000) 3302.
6. J.-H. Kim, S.I. Khartsev, and A.M. Grishin, *Appl. Phys. Lett.* 82, (2003) 4295.
7. S.P.S. Badwal, S.P. Jiang, J. Love, J. Nowotny, M. Rekas, and E.R. Vance, *Ceram. Int.*, 27 (2001) 419.
8. H.J. Hwang, A. Towata, M. Awano, and K. Maeda, *Scripta Materialia* 44 (2001) 2173.
9. C.H. Chen, H. Kruidhof, H.J.M. Bouwmeester, and A.J. Burggraaf, *Mater. Sci. Eng. B*, 39 (1996) 129.
10. S.J. Skinner, *Int. J. Inorg. Mater.*, 3 (2001) 113.
11. A.N. Petrov, O. F. Kononchuk, A.V. Andreev, V.A. Cherepanov, and P. Kofstad. *Solid State Ionics*, 80 (1995) 189.
12. A. Mineshige, M. Inaba, T. Yao, and Z. Z. Ogumi, K. Kikuchi, M. Kawase. *J. Solid State Chem.*, 121, (1996) 423.
13. J. Mizusaki, Y. Mima, S. Yamauchi, K. Fueki, H. Takagawa, *J. Solid State Chem.*, 80 (1989) 102.
14. P.M. Raccah, J.B. Goodenough, *J. Appl. Phys.*, 39 (1968) 1209. ledn.
15. M.A. Señarís-Rodríguez, M.P. Breijo, S. Castro, C. Rey, M. Sánchez, R.D. Sánchez, J. Mira, A. Fondado, and J. Rivas, *Int. J. Inorg. Mater.*, 1 (1999) 281.
16. Q. Huang, J. -M. Liu J. Li, H. C. Fang, H. P. Li, and C. K. Ong, *Appl. Phys. A*, (1999).
17. J.T. Cheung, P.E.D. Morgan, D.H. Lowndes, X-Y Zheng, and J. Breen, *Appl. Phys. Lett.* 62 (1993) 2045.
18. K.H. Wong, W. Wu, P.W. Chan, J.T. Cheung, *Thin Solid Films*, 312 (1998) 7.
19. S. Madhukar, S. Aggrawal, A.M. Dhote, R. Ramesh, A. Krishnan, D. Keeble, E. Poindexter, J. Appl. Phys., 81 (1997) 3543.
20. F. Wang and S. Leppävouri, J. Appl. Phys., 82 (1997) 1293.
21. M. Li, Z.-L. Wang, S. Fan, Q.-T. Zhao, G. Xiong, *Thin Solid Films*, 323 (1998) 304.
22. S.G. Ghonge, E. Goo, R. Ramesh, T. Sands, V.G. Keramidas, *Appl. Phys. Lett.*, 63 (1993) 1628.
23. M.A. Grishin, A.M. Grishin, S.I. Khartsev, U.O. Karlsson, *Integr. Ferroelectrics*, 39 (2001) 1301.
24. L. Fahua, X. Dingquan, P. Wenbin, W. Hongtao, Z. Jianguo, Z. Wen, *J. Kor. Phys. Soc.*, 32 (1998) S1471.
25. H.N. Sharaeef, B.A. Tuttle, W.L. Warren, D. Dimos, M.V. Raymond, M.A. Rodriguez, *Appl. Phys. Lett.*, 68 (1996) 272.
26. E. Bucher, W. Jantscher, A. Benisek, W. Preis, I. Rom, an F. Hofer, *Solid State Ionics*, 141-142 (2001) 375.
27. A. Pohl and G. Westin, *J. Am. Cer. Soc.*, 88 (2005) 2099.
28. A. Pohl, G. Westin, and K. Jansson, *Chem. Mater.*, 14, (2002) 1981.
29. Q.F. Chan, L.W. Helen, Q.Q. Zhang, C.L. Choy, *Ferroelectrics*, 232 (1999) 933.
30. K.-T. Kim, C.I. Kim, T.-H. Kim, Vacuum, 74 (2004) 671.
31. G. Westin, M. Ottosson, and A. Pohl, submitted to *Thin Solid Films*.
32. R. Bengtsson, S. Khartsev, A. Grishin, A. Pohl and G. Westin, *Thin Solid Films*, 467 (2004) 112.
33. R.W. Adams, E. Bishop, R.L. Martin, and G. Winter, *Aust. J. Chem.*, 19 (1966) 207.
34. N.Y. Turova, E.P. Turevskaya, V.G. Kessler, and M.I. Yanovskya, *The chemistry of metal alkoxides*, Kluwer Academic Publishers (2002).
35. B.P. Baranwal and R.C. Mehrotra, *Aust. J. Chem.*, 33 (1980) 37.
36. G. Westin, *Chem. Commun. (Stockholm University)* 4 (1994) 1-113.
37. G. Westin and M. Nygren, *J. Mater. Sci.*, 27 (1992) 1617.
38. I.A.M. Pohl, L.G. Westin and M. Kritikos, *Chem. Eur. J.* 7, (2001) 3438.
39. S.-Y. Bae, and S. X. Wang, *Appl. Phys. Lett.* 69. (1996) 121.
40. U. Hasenkox, C. Mitze, and R. Waser, *J. Am. Ceram. Soc.* 80, (1997) 2709.
41. C. Vázquez-Vázquez, M.C. Blanco, M.A. López-Quintela, R.D. Sáncez, J. Rivas, and S.B. Oseroff, *J. Mater. Chem.*, 8(4), (1998) 991.
42. S. V. Pietambaram, D. Kumar, R. K. Singh, and C. B. Lee, *Mat. Res. Soc. Symp.* 617, (2001) J3141.
43. K.-S. Hwang, H.-M. Lee, S.-S. Min, B.-A. Kang, *J. Sol–gel Sci. Technol*, 18 (2000) 175.

MICROPOROUS ZrO$_2$ MEMBRANE PREPARATION BY LIQUID-INJECTION MOCVD

S. Mathur, E. Hemmer, S. Barth, J. Altmayer and N. Donia
Leibniz Institute for New Materials, Nanocrystalline Materials and Thin Film Systems
Saarland University Campus, Building D2 2
D-66123 Saarbrücken, Germany

I. Kumakiri, N. Lecerf and R. Bredesen
SINTEF, Materials and Chemistry
PB 124 Blindern
N-0314 Oslo, Norway

ABSTRACT

Inorganic microporous membranes, presenting pores smaller than 2 nm, have an important potential for large-scale application in gas purification and separation. Given its good mechanical strength, thermal stability and chemical inertness, zirconia appears to be a promising material for the preparation of mesoporous and microporous membrane systems. We present here a membrane system based on the chemical vapor deposition of microporous zirconia on α-alumina / γ-alumina supports prepared by slip casting and sol-gel. For this purpose, Zr(OBu^t)$_4$ was chosen as precursor for the deposition of nanocrystalline ZrO$_2$ films. The thermal decomposition of the precursor in the liquid-injection CVD process was investigated as a function of the substrate temperature and different precursor injection frequencies. The as-deposited and post-annealed zirconia layers were investigated by EDX, XRD, SEM and AFM. First permporometry tests showed that the additional zirconia layer influenced the gas flux through the membrane in dependence of the CVD process parameters and resulted in a reduction of the total pore volume.

INTRODUCTION

Inorganic membranes, allowing a selective mass transport of species, have an important potential for industrial applications like purification and separation processes, particularly at high temperatures and under chemically harsh conditions [1].

Recently, microporous silica is the most investigated system for membrane applications. However, the silica system suffers from poor stability to humidity. The limited chemical resistance, especially in alkaline media or in strong electrolyte solutions, turns the choice of adequate ceramic materials out to be complex and restricted. Herein, zirconia appears to be a promising material for the preparation of mesoporous and microporous membrane systems due to its good mechanical strength, thermal stability and chemical inertness. Zirconia can in fact be used in aqueous filtration with much better alkali durability than other ceramic membranes such as alumina or silica [2].

Currently, slip-casting of suspended colloidal particles and sol-gel processes are the most common techniques for ceramic membrane preparation [3]. The obtained macroporous supports can be modified by chemical vapor infiltration to deposit solids within the pores. This strategy allows tuning the pore size distribution of the membranes [4]. In this study a membrane of microporous zirconia on α-alumina / γ-alumina supports was prepared by liquid-injection metal-organic chemical vapor deposition (LI-MOCVD).

For the deposition of ZrO_2 by MOCVD the range of suitable precursors includes β-diketonates [5, 6], fluorinated β-diketonates [6, 7] and alkoxides [8]. Although the alkoxide zirconium-*tert*-butoxide, $Zr(OBu^t)_4$, shows sufficient volatility to be suitable for decomposition and formation of ZrO_2 films in the conventional thermal MOCVD process [9], the use of a liquid-delivery system for the precursor implies several advantages: In liquid-injection MOCVD the solvent vapor is the major gas phase species and the concentration of the precursor in the gas phase is directly related to the concentration of the precursor in the liquid solution inducing an easy control of vapor pressure and precursor concentration in the gas flow. Furthermore, flow rate is fixed by the opening frequency and time of the piezovalve between vaporizer and reaction chamber offering the controlled supplies of precursor to reaction zone. This allows a simple control of film thickness by the number of injections. In addition, the precursor reservoir remains cold up to the point where the solution is vaporized providing any unwanted aging of the precursor and being a benefit under up-scalability and economical aspects [10].

EXPERIMENTAL

Membrane Preparation and Liquid-Injection MOCVD

Macroporous α-Al_2O_3 supports were fabricated via slip-casting, followed by firing at 1100 °C for 1 h and machining to the required dimension as well as polishing. Porosity of these flat supports was 35 % and they had an average pore size of 80 nm. In a following step, mesoporous γ-Al_2O_3 layers were deposited on the α-Al_2O_3 support by dip-coating under class-100 clean-room conditions. After dip-coating, the membranes were dried and fired in air at temperatures between 600 °C and 1000 °C for 3 h with a heating / cooling rate of 1 K / min.

The synthesis of zirconium *tert*-butyloxide ($Zr(OBu^t)_4$) followed the procedure described in literature [11] in a modified Schlenk type vacuum assembly, taking stringent precautions against atmospheric moisture. Deposition of thin ZrO_2 films on the γ-Al_2O_3 layer of the support was affected by a liquid-injection CVD process using a solution of 2 Vol % $Zr(OBu^t)_4$ in toluene. Precursor solution was injected for 30 min with 1 Hz, respectively 2 Hz injection frequency by 1 ms opening time of the piezovalve at 60 °C and 10^{-2} Torr reaction chamber pressure. The alumina support was inductively heated in a range of 500 to 600 °C using a graphite susceptor. Some samples were additionally annealed. Therefore, post-thermal treatment was affected at 500°C during 2 h under atmospheric conditions.

Instrumentation

Surface morphology was recorded on a scanning electron microscope JSM-6400F (JEOL). Surface roughness was analyzed by atomic force microscopy (TMX2000 Explorer AFM with standard Si_3N_4 tip). Phase characterization was done by X-ray diffraction (XRD) analysis at room temperature on a Siemens diffractometer D-5000 fitted with a SEIFERT X-ray tube operating with Cu K_α (λ = 0.154 18 nm).

Permporometry

Since the membranes prepared in this study are asymmetric, the BET method would not be adequate due to additional gas adsorption on the alumina support. Therefore, the mean pore diameter of the membranes has been estimated by means of permporometry based on ethanol / nitrogen permeation.

In this technique, gas permeation is measured by increasing vapor pressure from dry (no vapor) conditions to close to saturated vapor conditions. Capillary condensation occurs first in the smaller pores during the measurement and blocks the gas permeation. The simultaneous gas flux through the remaining open pores allows the measurement of the fraction of pores with pore size larger than the pores that are already blocked. Upon desorption the same process occurs in reverse order. Accordingly, gas permeation decreases with increasing vapor pressure in relation with the Kelvin condensation equation (1):

$$d_k = - 4\sigma \cos\theta \, / \, [\, V_m RT \, \ln(P/P_{sat})] \tag{1}$$

where d_k stands for the Kelvin radius, σ for the surface tension, θ for the contact angle, V_m for the molar volume and where P is the vapor pressure.

The pore size distribution of both γ-Al$_2$O$_3$ supports and zirconia membranes were examined. Ethanol was used as vapor. All the gas flux measurements were done at room temperature.

RESULTS AND DISCUSSION
Film Growth by LI-MOCVD and Characterization

The precursor Zr(OBu^t)$_4$ was chosen for its gas phase stability and for the presence of pre-formed Zr-O bonds which favor the formation of ZrO$_2$ at low processing temperatures. The use of a single molecular source simplifies the growth procedure and allows a precise control over phase composition and particle size in the deposits [12]. The aim of this study was to investigate the thermal decomposition of the molecular precursor as a function of the substrate temperature (500 °C, 500 °C followed by a 2 h annealing step at 500 °C and 600°C) and different precursor injection frequencies (1 Hz, 2 Hz).

Morphology of the as-deposited and annealed films was investigated in high-resolution scanning electron microscope. A ZrO$_2$ film deposited at 500 °C (1 Hz) showed a very smooth and dense surface morphology without visible defects (Fig. 1a). Due to higher precursor concentration in the reaction area at enhanced injection frequency (500°C, 2 Hz) grain growth prevails nucleation resulting in a heterogeneous morphology (Fig. 1b). In this case, film structure is built up by grains with facetted morphology with an average length of ca. 140 nm and an average width of ca. 60 nm. A similar effect with slightly smaller grains (about 120 nm in length and 45 nm in width) was obtained at 600 °C (1 Hz), however in this case nucleation of clusters that were unevenly distributed over the substrate was observed (Fig. 1c).

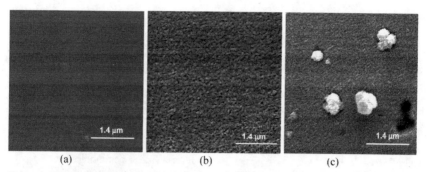

| (a) | (b) | (c) |

Figure 1: SEM micrographs of membranes deposited at 500°C, 1 Hz (a), 500°C, 2 Hz (b) and 600°C, 1 Hz (c): Development of the microstructure of the ZrO₂ membranes.

Investigations of surface roughness by atomic force microscopy confirmed the described development of the surface morphology. Films deposited at 500°C (1 Hz) showed only slightly increased roughness values independent of an additional post-annealing step (R_a = 4.6 nm, respectively 4.7 nm) when compared with the uncoated alumina support (R_a = 3.4 nm), while an increase in both injection frequency and substrate temperature led to enhanced roughness of about 6 nm (500°C, 2 Hz) and ca. 18 nm (600°C, 1 Hz), respectively.

X-ray diffraction profiles of the membranes deposited at different process parameters showed good crystallinity and indicated in all cases a mixture of tetragonal (t-phase) and monoclinic (m-phase) ZrO₂ (Fig. 2).

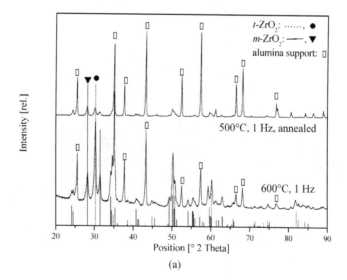

(a)

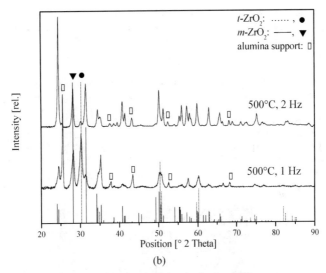

(b)

Figure 2: XRD patterns (Cu K$_\alpha$) of the ZrO$_2$ deposits obtained at different process parameters: (a) influence of temperature, (b) influence of injection frequency. Referent patterns: t-ZrO$_2$ [50-1089] (dotted line), m-ZrO$_2$ [37-1484] (drawn line).

The recorded patterns show an increase in contribution of m-ZrO$_2$ for higher injection frequency (Fig. 2b) and higher substrate temperature (Fig. 2a). This indicates a t-to-m transformation and / or grain growth mainly of the m-phase in dependence on temperature and precursor concentration in the reaction zone. The observed t-to-m transition at increased substrate temperature (600°C, 1 Hz, especially the reflexes at 2 θ = 34.160° and 34.383°, respectively 49.266° and 50.116° are indicating the m-phase) is conformal with the results of our former investigations for ZrO$_2$ film growth by thermal MOCVD which showed a phase transition already at 450 °C [9]. However, while the transition from t- to m-ZrO$_2$ was completed at 550 °C in case of thermal MOCVD, the tetragonal modification in the films prepared by liquid-injection MOCVD remained partially stable till 600 °C. Formation of the high temperature tetragonal phase (transition points of the bulk material: monoclinic to tetragonal: 1170 °C, tetragonal to cubic: 2370 °C) at low temperatures may be due to the lower surface energy of nanocrystalline t-zirconia. While Garvie [13, 14] reported about formation of stable t-ZrO$_2$ at low temperatures in case of particles smaller than the critical value of 10 nm in diameter and a t-to-m transformation of larger grains, Chraska et al. [15] confirmed that even larger grains of the t-phase could be stable in case of direct nucleation in the tetragonal modification. A similar effect is observed by increasing the injection frequency that is directly correlated with a higher precursor concentration in the gas phase inducing faster growth of the first nuclei on the substrate. Furthermore, average grain sizes increased slightly from ca. 12 nm to ca. 15 nm for t-phase and to ca. 19 nm for m-phase by doubling the injection frequency as calculated from the x-ray data using the Scherrer formula (reflexes used for calculations: ▼ and ●, Figure 2b). The different intensities of the reflexes for α-Al$_2$O$_3$ resulting from the alumina support are due to the enhanced film thickness in case of higher injection frequency. The post-annealing of the as deposited membranes at 500 °C provokes strong grain growth up to an average grain size of about 21 nm (reflexes used for calculations: ▼ and ●, Figure 2a). The strong α-Al$_2$O$_3$ signals in this case may indicate diffusion processes of Al^{3+} into the deposited zirconia layer.

Further, cross-section analysis of the membrane systems by high-resolution SEM was performed in order to investigate thickness of the zirconia membrane and microstructure within the membrane system. These investigations confirmed the prediction about enhanced film growth at higher injection frequency derived from XRD measurements and surface morphology. The thickness of the ZrO$_2$ layers is uniform for the entire coated surface. The thickness values are directly related to the process parameters. The growth rates at 500 °C were found to be 3 nm / min and 12 nm / min for injection frequency 1 and 2 Hz, respectively. Figure 3 illustrates the development of the pore size in the three layers of the membrane: The macroporous α-Al$_2$O$_3$ support, the mesoporous γ-Al$_2$O$_3$ interlayer and the zirconia top layer.

Figure 3: Cross-section SEM micrograph showing the three compounds of the membrane system (500°C, 1Hz): α-Al$_2$O$_3$ support base, γ-Al$_2$O$_3$ support coating and ZrO$_2$ top layer[16].

Flux Measurements and Pore Size Distribution

Flux measurements and estimation of the mean pore diameter of the membranes by means of permporometry based on ethanol / nitrogen permeation illustrated the effect of deposition process parameters on membrane behavior in gas permeation.

Gas flux in the γ-Al$_2$O$_3$ membrane varied in the range of 3 - 6 · 10^{-7} mol/m^2/s/Pa. Figure 4a presents the nitrogen flux measured through the membranes with increasing ethanol vapor pressure. Applying a ZrO$_2$ layer by LI-MOCVD decreases the permeation flux to 50 % to 10 %, depending on the CVD conditions, of the initial value of the pure γ-Al$_2$O$_3$ membrane. As expected, the smooth, dense films (500°C, 1 Hz) and the thick films with facetted structure (500°C, 2 Hz) provoked minor gas flux, while the membranes deposited at higher substrate temperatures (600°C, 1 Hz) suffering from pores and clusters induced enhanced gas flux. Further, a sharp flux drop at lower Kelvin radius (0.5 nm) was obtained for membranes deposited at increased temperature (600°C, 1 Hz) and after post-thermal treatment (500°C). In contrast, the membrane deposited at 500 °C with injection frequency of 1 Hz and 2 Hz showed no flux drop which is characteristic for denser layers. Most of the pores in γ-Al$_2$O$_3$ membrane are in the range of 1 to 2.7 nm (see open keys in Figure 4b). Please note that the pore size in the figure represents Kelvin radius which is smaller than the real pore size as the thickness of adsorption layer is not taken into account. The shape of pore size distribution curves was changed after applying ZrO$_2$ layer by CVD (see closed keys in Figure 4b). Approximately 50% of nitrogen permeation through CVD membranes was obtained through pores smaller than 1nm Kelvin diameter. This suggested that pore size was narrowed by CVD deposition. Some of the CVD membranes had pores larger than 3 nm. These large pores might be originated from support and got less influence by CVD deposition.

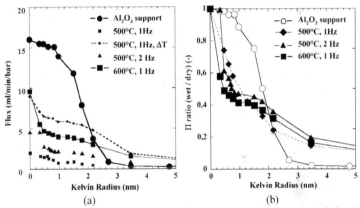

(a)   (b)

Figure 4: (a) Nitrogen flux with increasing ethanol vapor pressure for the uncoated alumina support and the ZrO$_2$ membranes deposited at different process parameters. (b) Pore size distribution (permeance ratio (wet flux / dry flux)) as a function of Kelvin radius.

CONCLUSIONS

Crack free ZrO$_2$ membranes have successfully been synthesized by gas phase deposition using Zr(OBu^t)$_4$ as precursor. Its thermal decomposition in the liquid-injection CVD led to a nanoporous ZrO$_2$ layer on the surface of the γ-Al$_2$O$_3$ support layer. Film characterization by X-ray diffraction method showed a mixture of t-ZrO$_2$ and m-ZrO$_2$ with further t-to-m transition at elevated substrate temperatures and injection frequencies. In analogy, changes in morphology from smooth and dense films to thick films with facetted microstructure were obtained.

Permporometry showed that the zirconia layer properties depend on the CVD process parameters. The developed procedure based on modification by pore size reduction of macroporous Al$_2$O$_3$ supports via zirconia deposition through gas phase technique presented the advantage to prepare zirconia membranes in one single step. While the sol-gel process, which usually necessitate clean room condition with careful drying and annealing, holds the high risk of agglomerations having automatically a negative effect on the pore size and the membrane quality this drawback does not appear when the membrane layer is deposited from the gas phase as it is achieved here.

This study hence emphasized the feasibility of this procedure for the preparation of zirconia membranes which is interesting from an academic and from an industrial point of view. Significant stability under milder steam conditions, as encountered in steam sterilization and pervaporation, is expected from the prepared membranes.

ACKNOWLEDGEMENTS

Authors are thankful to the German Academic Exchange Service (DAAD) and to the Norwegian Research Council (NFR) for providing financial assistance in the frame of the Project Based Personnel Exchange Programme (PPP) between Germany and Norway. N.L., I.K. and R.B. thank the NFR for the financial support provided in the NANOMAT Programme (Functional Oxides for Energy Technology. Part C). S.M., E.H., S.B., J.A. and N.D. would like to acknowledge the NATO-CLG Program (No. 979756) for supporting this work.

REFERENCES

[1]L. Cot, A. Ayral, J.J. Durand, C. Guizard, N. Hovnanian, A. Julbe and A. Larbot, "Inorganic membranes and solid state sciences", *Solid State Sci.*, **2**, 313-334 (2000).

[2]T. Yazawa, H. Tanaka, H. Nakamichi and T. Yokohama, "Preparation of water and alkali durable porous glass membrane coated on porous alumina tubing by sol-gel method", *J. Membr. Sci.*, **60**, 307-317 (1978).

[3]L. Cot, C. Guizard and A. Larbot, "Novel ceramic material for separation process: present and prospective applications in microfiltration and ultrafiltration", *Ind. Ceramic*, **8**, 143-148 (1988).

[4]C.F. Lin, D.L. Flowers and P.K.T. Liu, "Characterization of ceramic membranes II. Modified commercial membranes with pore size under 40 Å", *J. Membr. Sci.*, **92**, 45-58 (1994).

[5]J. Si and S.B. Desu, C.Y. Tsai, "Metal-organic chemical vapor deposition of ZrO$_2$ films using Zr(thd)$_4$ as precursors", *J. Mater. Res.*, **9**, 1721-1727 (1994).

[6]M. Balog, M. Schieber, S. Patai and M. Michman, "Thin films of metal oxides on silicon by chemical vapor deposition with organometallic compounds. I", *J. Cryst. Growth*, **17**, 298-301 (1972).

[7]C.S. Hwang and H.J. Kim, "Deposition and characterization of ZrO$_2$ thin-films on silicon substrate by MOCVD" *J. Mater. Res.*, **8**, 1361-1367 (1993).

[8]Z. Xue, B.A. Vaartstra, K.G. Caulton, M.H. Chisholm and D.L. Jones, "Chemical vapor-deposition of cubic-zirconia thin-films from zirconium alkoxide complexes", *Eur. J. Solid State Inorg.Chem.*, **29**, 213-225 (1992).

[9]S. Mathur, J. Altmayer and H. Shen, "Nanostructured ZrO$_2$ and Zr-C-N coatings from chemical vapor deposition of metal-organic precursors" *Z. Anorg. Allg. Chem.*, **630**, 2042-2048 (2004).

[10]P.J. Wright, M.J. Crosbie, P.A. Lane, D.J. Williams, A.C. Jones, T.J. Leedham and H.O. Davies, "Metal organic chemical vapor deposition (MOCVD) of oxides and ferroelectric materials", *J. Mater. Sci.: Materials in Electronics*, **13**, 671-678 (2002).

[11]I.M. Thomas, "The preparation of alkoxides and triethylsilanolates of Ti, Zr, V, Nb, Ta and Sn from the dialkylamides", *Can. J. Chem.*, **39**, 1386-1388 (1961).
[12]S. Mathur and H. Shen, "Inorganic nanomaterials from molecular templates" in *"Encyclopedia of nanoscience and nanotechnology"*,H. Nalwa (Ed.), Amercican Scientific Publisher, **4**, 131-191 (2004).
[13]R.C. Garvie, "The occurrence of metastable tetragonal zirconia as a crystallite size effect" *J. Phys. Chem. Solids*, **69**, 1238-1243 (1965).
[14]R.C. Garvie, "Stabilization of tetragonal structure in zirconia microcrystals" *J. Phys. Chem. Solids*, **82**, 218-224 (1978).
[15]T. Chraska, A.H. King and C.C. Berndt, "On the size-dependent phase transformation in nanoparticulate zirconia" *Mater. Sci. Eng. A*, **286**, 169-178 (2000).
[16]S. Mathur, E. Hemmer, S. Barth, J. Altmayer, N. Donia, N. Lecerf, I. Kumakiri and R. Bredesen, "Microporous ZrO$_2$ film preparation by chemical vapor deposition", *Proc. 9th Int. Conf. on Inorganic Membranes* / ed. R. Bredesen and H. Raeder, Lillehammer - Norway, 524-527 (2006).

PHOTODECOMPOSITION OF ACETONE ON ZrO_x-TiO_2 THIN FILMS IN O_2 EXCESS AND DEFICIT CONDITIONS

L. Österlund* and A. Mattsson
FOI CBRN Defence and Security
SE-901 82 Umeå, Sweden.
*) Email: lars.osterlund@foi.se

M. Leideborg and G. Westin
Department of Materials Chemistry, Ångström Laboratory, Uppsala University
SE-751-21 Uppsala, Sweden

ABSTRACT
We present a molecular spectroscopy study of the photo-induced decomposition of acetone on zirconium oxide (1-5 at.% Zr) modified TiO_2 thin films with anatase modification. A detailed account of the decomposition products and reaction pathways is presented. It is found that surface coordinated acetate, formate, formaldehyde and carbonate form in the course of the photoreaction. The preferred coordination is bridging bidentate (μ-coordination). The ZrO_x addition leads to a decreased decomposition rate in synthetic air (O_2 excess) and the results in synthetic air are interpreted in terms of an increased recombination rate in the ZrO_x:TiO_2 films, rather than differences in reaction pathways. In contrast, in O_2 free environment employing N_2 gas, the photoreaction involving lattice O leads to a different decomposition pathway than when O_2 is present in excess. A striking result is that the ZrO_x:TiO_2 films exhibit an exceptional stable activity over time and usage; much better than the pure TiO_2 films. Thus despite an inferior activity the sustain activity of the $ZrOx$-TiO_2 films suggests means to circumvent deactivation of photocatalyst materials.

INTRODUCTION
Titanium dioxide has for the last decades been a promising photocatalyst used for air and water purification.[1,2] The wide band gap of the material necessitates UV-light (λ<400 nm) for photocatalytic activity limiting its use. Reducing the band gap by doping TiO_2 with different materials[3-5] is a possible way to reduce the band gap of TiO_2, and hence increase the efficiency of the photocatalyst in solar light applications. Large attention are now also devoted to other types of metal oxides than TiO_2 and mixed metal oxides.[6-10] However, mixing TiO_2 with metal oxides or using mixed phase metal oxides systems may introduce unwanted effects. This may lead to more recombination centres, which decrease the life time of the excited electron-hole pairs. It may also lower the conduction band below the O_2 affinity level and thereby reduce the formation of oxygen radicals, which are a key intermediate in the photocatalytic process[11] that altogether decreases the photocatalytic performance of the material. Mixed metal oxide systems may also lead to phase separation or surface segregation with regions exhibiting different physico-chemical properties. Thus the adsorption and photochemical properties may vary within the same nanoparticle.

In this paper we report results concerning the photo-induced decomposition of acetone on nanostructured TiO_2 thin films doped with different concentrations of Zirconium oxide (ZrO_2). TiO_2-ZrO_2 nanostructured materials have been used in dye-sensitive solar cell with good results.[12] Preliminary studies performed by two of us (GW and ML) of ZrO_2 doped TiO_2 in wet solar cells, indicated an enhanced efficiency at low Zr concentrations. In this study we explore whether these results also imply improved organic photodegradation efficiency for solar light degradation.

EXPERIMENTAL SECTION

Materials

Zirconium oxide doped anatase TiO$_2$ thin films were made by a variation of the sol-gel method described elsewhere.[13] The resulting materials consisted of TiO$_2$ nanoparticles with anatase modification (<d>=25 nm) with 0, 1, 2 and 5 at.% Zr. We use the nomenclature ZrO$_x$:TiO$_2$ for the materials throughout this paper since the exact stoichiometry and structure of the mixed oxide have not been determined. Thin films of ZrO$_x$:TiO$_2$ were spin coated on CaF$_2$ substrates. The structural, optical and chemical properties of the synthesized materials were characterized by a range of different techniques described in detail elsewhere.[13,14] Scanning electron microscope images were obtained with a FEG-SEM Leo 1550 Gemini instrument. Transmission electron microscopy was done with a Jeol 2000 FXII instrument. Reflectance micro-Raman spectra of samples heat treated in air for 30 min at 723 K were recorded with a Renishaw 2000 spectrometer using a 783 nm laser diode light source. Grazing incidence X-ray diffractograms were obtained for samples heat treated at 723 K with a Siemens D-5000 instrument. UV-Vis measurements for the films were made with a Perkin Elmer Lambda 18 UV-VIS Spectrometer.

Figure 1. SEM image of ZrO$_x$:TiO$_2$ with 2 at.% Zr.

Photodegradation experiments

The thin films were irradiated with simulated solar light generated by a Xe arc lamp source operated at 200W employing air mass filters AM1.5 described elsewhere.[13] The photon power on the sample was determined to be 166 mW cm^{-2} for wavelengths between 200 and 800 nm, corresponding to ca 12 mW cm^{-2} for $\lambda < 390$ nm.

Fourier transform infrared (FTIR) transmission measurements were made in a vacuum pumped FTIR spectrometer equipped with a vacuum tight transmission reaction cell, which allowed for simultaneous UV illumination, mass spectrometry and in situ FTIR in controlled atmosphere.[13] Repeated FTIR spectra were measured with 4 cm^{-1} resolution and 135 scans and 30 sec dwell time between consecutive spectra. The sample was held at 299 K and two sets of experiments were done employing either a 100 ml min^{-1} gas flow of synthetic air (20% O$_2$ and

80% N$_2$) or a 100 ml min^{-1} pure N$_2$ gas flow through the reaction cell. Prior the each measurement the samples were annealed at 673 K in synthetic air and subsequently cooled to 299 K in the same feed. For experiments without O$_2$ in the gas feed, the sample was purged in 100% N$_2$ for ca 30 minutes prior to acetone dosing. Acetone (Analytical Grade, Scharlau) was added to the gas feed through a homebuilt gas generator. The independently calibrated acetone injection in the gas feed obtained with these parameters was 0.14 ± 0.02 mg min^{-1}, which corresponds to an ideal steady state concentration of 595 ± 110 ppm in the feed. No other gas species than acetone was detected mass spectrometrically with this gas evaporation set-up. In each experiment the sample was exposed to 15 minutes of acetone unless otherwise stated. After acetone dosing the sample was kept in the gas feed for ca 15 minutes prior to illumination. FTIR background was collected on a clean sample during 1 minute (265 scans) in synthetic air or N$_2$ feed at 299K.

RESULTS AND DISCUSSIONS
UV-Vis absorption
    In semiconductors the absorption coefficient α for indirect optical transitions at energies $E=h\nu$ is well described by $\alpha h\nu = (h\nu - E_g)^2$, where E_g is the optical band edge. In Figure 2 is shown a plot of $(Ah\nu)^{0.5}$ vs. λ, where A is the absorbance as measured by UV-Vis spectrometry. The linear shape of the curves just above optical absorption edge (the lowest lying optical transition) shows that the all ZrO$_x$:TiO$_2$ films are indirect band gap semiconductors, independent of doping concentrations less than 5 at.% Zr. The optical absorption edge shifts either up or down depending on Zr concentration. At 1% Zr the band gap shifts slightly down $\Delta E_g \approx -0.02$ eV and up $\Delta E_g \approx 0.02$ eV for 2% Zr, while it shifts up $\Delta E_g \approx 0.11$ with 5% Zr compared to pure anatase TiO$_2$.

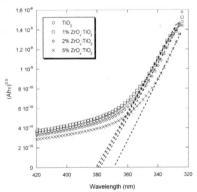

Figure 2. Plot of $(Ah\nu)^{0.5}$ vs. λ for 1, 2, and 5 % ZrO$_x$:TiO$_2$ and pure anatase TiO$_2$.

Acetone adsorption
    The bottom spectra in Figures 3-5 and Figure 6-8 show the FTIR spectra obtained after acetone adsorption (prior to illumination) in synthetic air and nitrogen gas on TiO$_2$, 2% ZrO$_x$:TiO$_2$ and 5% ZrO$_x$:TiO$_2$, respectively. The vibrational frequencies are similar to reported data for condensed acetone.[15] In particular the v(C=O) peak, which is sensitive to the local

chemical environment and acetone coverage[13], remains approximately constant. It is seen that neither doping concentration of Zr or feed gas (synthetic air or N_2) affects the vibrational properties of adsorbed acetone.

Photodecomposition of acetone in synthetic air

Figures 3-5 show FTIR spectra obtained on ZrO_x:TiO_2 with varying Zr doping concentrations at different illumination times. The new absorption bands that appear in the spectra represent intermediate species formed on the particles surfaces upon acetone photodecomposition. Due to differences in the decomposition rate depending on Zr concentration, the surface species formed in the course of the photoreaction appear at different times (and disappear after extended illumination). It can be inferred that the acetone decomposition rate decreases with increasing Zr doping (see Table 2 below). This is also true for the intermediate species.

Inspecting the FTIR spectra more closely a number of detailed conclusions can be made. The vibrational frequencies and peak assignments from this detailed assessment are shown in Table 1. In the region between 1500-1590 cm^{-1} where $v_a(COO)$ bands due to formate, acetate and bicarbonate species are expected to appear,[16-20] it is seen that with increasing Zr concentration the relative intensity of a broad, asymmetric ~1552 cm^{-1} peak becomes more pronounced compared to the peak at ~1586 cm^{-1}. This latter peak is thus very likely composed of several bands. Furthermore, the absorption band at ~1441 cm^{-1} becomes stronger with increasing Zr concentration. On the pure TiO_2 sample this peak is fairly small, whereas it is one of the largest peaks on the 5% ZrO_x:TiO_2 film. Within 10 minutes it is one of the largest peaks although the overall process is slower on this film. Based on these observations and the measured slower degradation rate on the high Zr loaded samples, we assigned the ~1441 cm^{-1} mode to the $v_s(COO)$ mode in acetate,[20] and the ~1552 cm^{-1} peak to overlapping $v_a(COO)$ modes in bridge-bonded formate (μ-format) and acetate. The increasing intensity of the ~1552 cm^{-1} peak with increasing Zr concentration is thus mainly due to acetate. The ~1586 cm^{-1} peak is due to $HCOO^-$ ions.[13,21] Surface coordinated carbonates species (μ- and η^1-carbonate) form after extended illumination, probably due to reactions with adsorbed CO_2 and O_2 (see Table 1).[19]

In the high wavenumber region the characteristic $v(CHO)$ peak from formaldehyde[18] is seen around ~2741 cm^{-1} and in the fingerprint region the peak at ~1740 cm^{-1} follows the same time-evolution and is therefore also attributed to formaldehyde. Other peaks not attributed to acetone in the $v(CH)$ spectral region is found at ~2848 cm^{-1}, ~2865 cm^{-1} and at ~2951 cm^{-1}. The latter peak is initially obscured by the $v_s(C-H)$ acetone peak, but it becomes more pronounced as the decomposition of acetone proceeds. Comparing the time-scale for the peaks it is evident that they all correlate to the formate and acetate peaks found in the fingerprint region. In the same way as the acetate concentration increases with increasing Zr concentration, the same trend can be seen also for the ~2846 cm^{-1} and ~2865 cm^{-1} peaks, where the latter increases with increasing Zr concentration. Thus the ~2846 cm^{-1} and ~2865 cm^{-1} peaks are assigned to the $v(CH)$ stretching mode in $HCOO^-$ and acetate, respectively. The ~2951 cm^{-1} peak seen on all samples is attributed to the well-known $v_a(COO)$+ $\delta(CH)$ combination band in μ-formate.[18] These results suggest that the trend of decreasing photoreaction rate is not due differences in the decomposition pathways for the specific organic adsorbate, and/or bonding to specific sites on ZrO_x:TiO_2 since the main absorbance bands appear at similar frequency, irrespective of Zr concentration, albeit with different relative absorbance. Thus we conclude that the reduced photodecomposition rate upon Zr doping instead is most likely due to an enhanced recombination rate of the electron-hole pairs and/or reduced interfacial radical formation rates. One possibility is that Zr introduces additional exciton traps in the form of lattice defects.

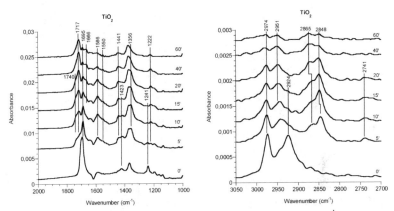

Figure 3. FTIR spectra of TiO$_2$ thin films in synthetic air in the 1000-2000 cm^{-1} and 2600-3050 cm^{-1} regions respectively, prior to (t=0 min) and during UV illumination.

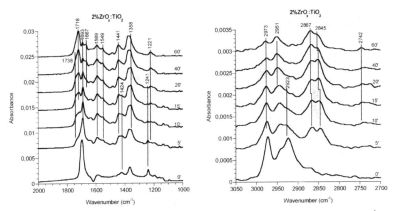

Figure 4. FTIR spectra of 2% ZrO$_x$:TiO$_2$ thin films in synthetic air in the 1000-2000 cm^{-1} and 2600-3050 cm^{-1} regions respectively, prior to (t=0 min) and during UV illumination.

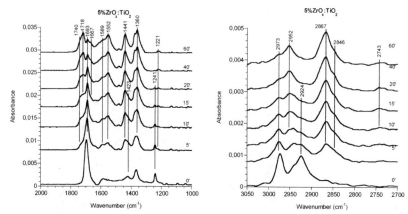

Figure 5. FTIR spectra of 5% ZrO_x:TiO_2 thin films in synthetic air in the 1000-2000 cm⁻¹ and 2600-3050 cm⁻¹ regions respectively, prior to (t=0 min) and during UV illumination.

Table 1. Vibrational frequency and suggested peak assignment for selected absorption bands detected on the pure and 5% Zr doped TiO_2 thin films in the course of the photoreaction. The compilation was made based on the following studies: Refs. [13,17,18,20-24]

Wavenumber (cm⁻¹)	Molecule	Band assignment	Proposed structure
1025	Methoxy	$v_s(CO)$	
1128	Methoxy	$v_a(CO)$	
1222	Carbonate	$v(CO)$	Bridged
1356	Formate	$v_s(COO)$	μ-formate
1372	Formate	$\delta(CH)$	
1441	Acetate	$v_s(COO)$	Ion
1540-1560*	Formate Acetate	$v_a(COO)$	μ-formate μ-acetate
1586	Formate	$v_a(COO)$	Ion
1666	Carbonate	$v(CO)$	Bridged
1717	Carbonate	$v(C=O)$	Monodentate
1740	Aldehyde	$v(C=O)$	
2741	Aldehyde	$v(CHO)$	
2819	Methoxy	$v_s(CH)$	
2848	Formate	$v(CH)$	Ion
2865	Acetate	$v(CH)$	
2940	Methoxy	$v_a(CH)$	
2951	Formate	$v_a(COO)+\delta(CH)$	

*Overlapping peaks

Photodecomposition in nitrogen gas

The FTIR spectra for the different ZrO_x:TiO_2 thin films in O_2 deficient conditions (N_2 gas) in the region between 1000-2000 cm^{-1} and 2600-3050 cm^{-1} are shown in Figures 6-8. The photodecomposition without O_2 in the gas feed is significantly slower (see Table 2 below), since it in this case is limited by lattice oxygen diffusion, which is the rate limiting step at small O_2 concentrations or O_2-free environments.[25,26] This means that many FTIR absorption bands due to intermediates are not seen in Figure 6-8, since they have not evolved within the time frame of the experiments. However, the FTIR data yield qualitatively different results compared to the photodecomposition experiments in synthetic air. Remarkably, the acetone photodecomposition rate increases with Zr doping. Consequently, the carbonate peaks are only seen as small peaks on the films with 2% and 5% Zr (see Table 1), which have the highest photodecomposition rate. This is also true for the aldehyde peaks.

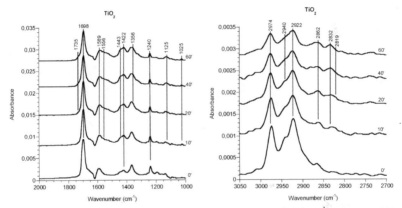

Figure 6. FTIR spectra of TiO_2 thin films in N_2 gas in the 1000-2000 cm^{-1} and 2600-3050 cm^{-1} regions respectively, prior to (t=0 min) and during UV illumination.

An analogous spectral analysis as presented above for the photodecomposition in synthetic air reveals that also in oxygen-free environment acetone decomposes into surface bound acetate, which is further oxidized to formate. However, a main difference in N_2 is that significant amounts of methoxy species form on the surface. This suggests a different total oxidation pathway than in synthetic air. These findings are in qualitative good agreement with the proposed mechanism by Muggli and Falconer.[25] They reported that the pathway for photo-induced decomposition of acetic acid is different depending on whether O_2 or lattice O is involved. In the pathway involving O_2, the α-carbon in acetic acid forms CO_2 and the methyl group produces methane. In the pathway involving lattice O, oxygen is extracted from the TiO_2 lattice and the α-carbon forms CO_2 and two methyl groups form ethane. Thus, in the latter pathway more methoxy species are expected to form. These results are supported by the relatively smaller amount of carbonate formation in the O_2-free case, since the carbonate formation pathway through reactions of adsorbed CO_2 and O_2 is constricted. The formation of acetate is clearly evident from Figure 6-8, where the ~1552 cm^{-1} peak grow as a function of illumination time, and appears significantly faster on films with high Zr loading. Similarly, in the v(CH) region the ~2846 cm^{-1} peak due to formate is missing (or is very weak), while the ~2865 cm^{-1} peak due to acetate is present. With increasing Zr concentration it is evident that new peaks at ~2817 cm^{-1} and ~2940 cm^{-1} appear in the spectra. The latter two are attributed to the v$_s$(CH) and v$_s$(CH) modes in methoxy in good agreement with previous studies.[24]

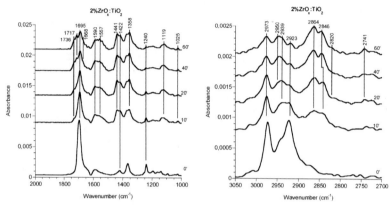

Figure 7. FTIR spectra of 2% ZrO$_x$:TiO$_2$ thin films in N$_2$ gas in the 1000-2000 cm^{-1} and 2600-3050 cm^{-1} regions respectively, prior to (t=0 min) and during UV illumination.

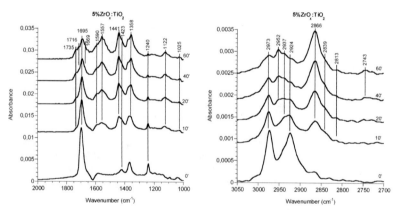

Figure 8. FTIR spectra of 5% ZrO$_x$:TiO$_2$ thin films in N$_2$ gas in the 1000-2000 cm^{-1} and 2600-3050 cm^{-1} regions respectively, prior to (t=0 min) and during UV illumination.

Photodecomposition rate

The photodecomposition rate constant was determined using the $v(C-C)$ acetone peak at ~1241 cm^{-1} and not the strong acetone $v(C=O)$ peak, since during the decomposition intermediate species with overlapping vibrational energies evolve. This problem is avoided by using the $v(C-C)$ peak. Assuming that the acetone photodecomposition is a pseudo-first order process, then $A(t)=A_0 e^{-kt}$, where $A(t)$ is the integrated $v(C-C)$ peak area at time t, A_0 the initial absorbance and k is the photodecomposition rate constant. The rate constants for the different measurements are given in Table 2 below. It is clearly seen that in an O_2-rich environment pure anatase TiO_2 is a better photocatalyst compared to the ZrO_x:TiO_2, which exhibit a decreasing activity with increasing Zr concentration. We have also performed measurements with ZrO_x:TiO_2 films on Si substrates, which also shows a lower activity with increasing Zr doping. In an O_2-free environment the results are reversed: The pure anatase TiO_2 films is inferior to the ZrO_x:TiO_2 films, which show an increasing activity with increasing Zr concentrations. However the measurements on silicon substrates in O_2 deficit conditions showed a significantly lower activity compared to the ones given here, which makes it difficult to discern any trends whether Zr doping is favourable in O_2-free environments and why. The measurements on Si substrates were also made with a lower annealing temperature, 573K instead of 673K, which could be insufficient to oxidize all organics on the surface and thereby further complicate the comparison.

Table 2. Acetone photodecomposition rate constants (min^{-1}) for the different ZrO_x:TiO_2 films in synthetic air ("O_2") and nitrogen gas ("N_2").

	Rate constant (min^{-1}) in O_2	Rate constant (min^{-1}) in N_2
TiO_2	-0,2904	-0,0138
1% Zr	-0,1512	-0,0424
2% Zr	-0,1398	-0,0790
5% Zr	-0,1056	-0,0648

Deactivation

Repeated exposures to acetone and subsequent cleaning at elevated temperatures and measurements in synthetic air have been done after a different number of exposures and cleaning cycles. In Table 3 are shown the rate constants determined in the same manner as described in the previous section. The number in parenthesis depicts the number of times the samples have been cleaned and exposed to acetone.

The trends discernable in Table 3 become clear when the rate constants are plotted against the number of exposure/annealing cycles (Figure 9). It can be seen that for the pure TiO_2 films the activity drops with the number of exposure/annealing cycles. Interestingly, on the films with Zr no such decline of the photodecomposition rate is observed. The reason for the deactivation of TiO_2 can be due to a number of different things such as healing, blocking or encapsulation of reactive sites (such as vacancies). Also annealing can increase the particle size (and reduce the porosity) and reduce number of reactive active edge or grain boundary sites. In any case it is clear that this is an irreversible effect which cannot be replenished by simple annealing in O_2.

In contrast, the ZrO$_x$:TiO$_2$ films maintain their activity during repeated exposure/annealing cycles. In analogy with previous reports on ZrO$_2$-TiO$_2$ mesoporous materials,[10] we tentatively attribute this to a structural effect, whereby Zr stabilizes the particle structure, which arguable may preserve active sites and/or counteract particle growth. This is of utmost importance in practical applications and suggests means to balance good photocatalytic activity with sustained operability.

Table 3. Rate constants for the decay of the acetone v(C-C) peak at ~1241 cm^{-1} in synthetic air for the different ZrO$_x$:TiO$_2$ films. The number of exposure/annealing cycles is indicated in parenthesis, along with the annealing temperatures.

	ZrO$_x$:TiO$_2$ on CaF$_2$	
	k (min^{-1})	T_{anneal}
TiO$_2$	-0,4878 (2)	673 K
	-0,3946 (3)	673 K
	-0,2904 (5)	673 K
1% Zr	-0,1549 (1)	673 K
	-0,1491 (3)	673 K
	-0,1511 (4)	673 K
	-0,1512 (5)	673 K
2% Zr	-0,1935 (1)	673 K
	-0,1086 (3)	673 K
	-0,1488 (4)	673 K
	-0,1398 (5)	673 K
5% Zr	-0,1266 (1)	673 K
	-0,0842 (3)	673 K
	-0,1056 (5)	673 K

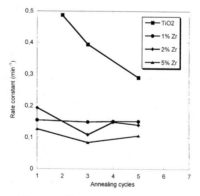

Figure 9. The photodecomposition rate of acetone on anatase TiO$_2$ and ZrO$_x$:TiO$_2$ annealed at 673K versus number of exposures and annealing cycles.

CONCLUSIONS

We have shown that ZrO_x-TiO_2 thin films with anatase modification made by sol-gel methods yield materials with inferior acetone photodegradation rate compared to pure anatase TiO_2 films. The decomposition products are similar on all materials in O_2 excess and exhibit the same vibrational energy losses in FTIR, with small or no chemical shifts. The inferior photodegradation rate for the ZrO_x:TiO_2 films is therefore tentatively attributed to an increased exciton trapping, rather than to differences in the reaction pathways. However, in O_2-free environment, a new reaction pathway involving lattice O becomes important which yields significant amount of methoxy groups. Interestingly, the sustained activity after several exposure/annealing cycles is improved by ZrO_x. This is of utmost importance in practical applications and suggests means to balance good photocatalytic activity with sustained operability.

REFERENCES

[1]A. Fujishima, K. Hashimoto and T. Watanabe, TiO_2 Photocatalysis. Fundamentals and Applications, BKC: Tokyo (1999)

[2]J. Peral, X. Domenech and D. F. Ollis, Heterogeneous photocatalysis for purification, decontamination and deodorization of air, *J. Chem. Technol. Biotechnol.*, **70**, 117-140 (1997)

[3]M. Anpo, Preparation, Characterization, and Reactivities of Highly Functinal Titanium Oxide-Based Photocatalysts Able to Operate under UV-Visible Light Irradiation: Approaches in Realizing High Efficiency in the Use of Visible Light, *Bull. Chem. Soc. Jpn.*, **77**, 1427-1442 (2004)

[4]R. Asahi, T. Morikawa, T. Ohwaki, K. Aoki and Y. Taga, Visible-Light Photocatalysis in Nitrogen-Doped Titanium Oxides, *Science*, **293**, 269-271 (2001)

[5]T. Umebayashi, T. Yamaki, H. Itoh and K. Asai, Band Gap Narrowing of Titanium Dixode by Sulfur Doping, *Appl. Phys. Lett.*, **81**, 454-456 (2002)

[6]I. P. Parkin and R. G. Palgrave, Self-Cleaning Coatings, *J. of Mat. Chem*, **15**, 1689-1695 (2005)

[7]M. Anpo, Photocatalysis on titanium oxide catalysts, *Catalysis Surveys*, **1**, 169-179 (1997)

[8]K. Y. Jung and P. B. P., Photoactivity of SiO_2/TiO_2 and ZrO_2/TiO_2 Mixed Oxides Prepared by Sol-Gel Method, *Mater. Lett.*, **58**, 2897-2900 (2004)

[9]T. Tatsuma, S. Saitoh, P. Ngaotrakanwiwat, Y. Ohko and A. Fujishima, Energy Storage of $TiO2$-$WO3$ Photocatalysis Systems in the Gas Phase, *Langmuir*, **18**, 7777-7779 (2002)

[10]X. Wang, J. C. Yu, Y. Chen, L. Wu and X. Fu, ZrO_2-Modified Mesoporous Nanocrystalline TiO_2-xNx as Efficient Visible Light Photocatalyst, *Environ. Sci. Technol.*, **40**, 2369-2374 (2006)

[11]*Photocatalytic Purification and Treatment of Water and Air,* Eds. D. F. Ollis and H. Al-Ekabi, Elsevier, Amsterdam, 1993

[12]A. Kitiyanan, S. Ngamsinlapasathian, S. Pavasupree and S. Yoshikawa, The Preparation and Characterization of Nanostructured TiO_2-ZrO_2 Mixed Oxide Electrode for Efficient Dye-Sensitized Solar Cells, *J. Solid State Chem.*, **178**, 1044-1048 (2005)

[13]A. Mattsson, M. Leideborg, K. Larsson, G. Westin and L. Österlund, Adsorption and Solar Light Decomposition of Acetone on Anatase TiO_2 and Niobium Doped TiO_2 Thin Films, *J. Phys. Chem. B*, **110**, 1210-1220 (2006)

[14]M. Leideborg and G. Westin, Preparation of Ti-Nb-O Nano Powders and Studies of the Structural Development on Heat-Treatment,*Ceramics: Getting into the 2000's - Part B 9th CIMTEC-World Ceramics Congress and Forum on New Materials*, Florence 1998.

[15]S. E. Stein, Infrared Spectra, *NIST Chemistry WebBook, NIST Standard Reference Database Number 69*, Eds. P. J. Linstrom and W. G. Mallard, June 2005, National Institute of Standards and Technology, Gaithersburg, MD, 20899 (http://webbok.nist.gov)

[16]L. Österlund and A. Mattsson, Surface Characteristics and Electronic Structure of Photocatalytic Reactions on TiO_2 and doped TiO_2 Nanoparticles, *Proc. of SPIE*, **6340**, 634003-1 (2006)

[17]G. Busca and V. Lorenzelli, Infrared Spectroscopic Identification of Species Arising from Reactive Adsorption of Carbon Oxides on Metal Oxide Surfaces, *Mater. Chem.*, **7**, 89-126 (1982)

[18]K. Nakamoto, Infrared and Raman Spectra of Inorganic and Coordination Compounds, John Wiley & Sons: New York (1997)

[19]J. R. S. Brownson, I. M. Tejedor-Tejedor and M. A. Anderson, Photoreactive Anatase Consolidation Characterized by FTIR Spectroscopy, *Chem. Mater.*, **17**, 6304-6310 (2005)

[20]J. M. Coronado, S. Kataoka, I. Tejedo-Tejedor and M. A. Anderson, Dynamic phenomena during the photocatalytic oxidation of ethanol and acetone over nanocrystalline TiO2: simultaneous FTIR analysis of gas and surface species, *J. Catal.*, **219**, 219-230 (2003)

[21]F. P. Rotzinger, J. M. Kesselman-Truttman, S. J. Hug, V. Shklover and M. Grätzel, Structure and Vibrational Spectrum of Formate and Acetate Adsorbed from Aqueous Solution onte the TiO_2 Rutile (110) Surface, *J. Phys. Chem. B*, **108**, 5004-5017 (2004)

[22]L.-F. Liao, W.-C. Wu, C.-Y. Chen and J.-L. Lin, Photooxidation of Formic Acid vs Formate and Ethanol vs Ethoxy on TiO_2 and Effect of Adsorbed Water on the Rates of Formate and Formic Acid Photooxidation, *J. Phys. Chem.*, **105**, 7678-7685 (2001)

[23]B. E. Hayden, A. King and M. A. Newton, Fourier Transform Reflection-Absorption IR Spectroscopy Study of Formate Adsorption on TiO_2(110), *J. Phys. Chem. B*, **103**, 203-208 (1999)

[24]C. Rusu and J. T. Yates, Jr., Adsorption and Decomposition of Dimethyl Methylphosphonate on TiO_2, *J. Phys. Chem. B.*, **104**, 12292-12298 (2000)

[25]D. S. Muggli and J. L. Falconer, Parallel Pathways for Photocatalytic Decomposition of Acetic Acid on TiO_2, *J. Catal.*, **187**, 230-237 (1999)

[26]D. S. Muggli and J. L. Falconer, Role of Lattice Oxygen in Photocatalytic Oxidation on TiO_2, *J. Catal.*, **191**, 318-325 (2000)

DESIGN, FABRICATION AND ELECTRONIC STRUCTURE OF ORIENTED METAL OXIDE NANOROD-ARRAYS

Lionel Vayssieres
National Institute for Materials Science, International Center for Young Scientists
Tsukuba, Ibaraki, Japan 305-0044

ABSTRACT

Materials chemistry has emerged as one of the most consistent fabrication tool for the rational delivery of high purity functional nanomaterials, engineered from molecular to macroscopic scale at low cost and large scale. An overview of the major achievements and latest advances of a recently developed growth concept and low temperature aqueous synthesis method for the fabrication of purpose-built large bandgap metal oxide semiconductor materials and oriented nano-arrays are presented. Important fundamental insights of direct relevance for semiconductor technology, optoelectronics, photovoltaics, and solar hydrogen generation are revealed by in-depth investigations of the electronic structure of metal oxide nanostructures with new morphology and architecture carried out at synchrotron radiation facilities.

INTRODUCTION

The controlled fabrication of well-defined and well-ordered one-dimensional large bandgap semiconductor nanomaterials such as nanorods, nanowires and nanobelts as well as their large scale manufacturing at low cost remain is of crucial importance to unfold the exciting and promising future of nanodevices. In addition to the rational and economical manufacturing of nanostructures, a better fundamental knowledge of their electronic structure as well as their physical, chemical, electrical and interfacial and structural properties is necessary to fully reveal and exploit their fascinating potentials. To achieve such ambitious targets, a new approach for the rational development of semiconductor nanostructures has been developed. A strategy based on the chemical and electrostatic lowering of the water-oxide interfacial energy of the systems[1] allowed a direct thin film growth by heteronucleation onto various substrates (amorphous, poly-, or single-crystalline) of large physical areas. Such approach has the capability to generate functional nanomaterials at large scale low cost and low temperature. For instance, crystalline metal oxide nanostructures consisting of oriented multi-dimensional arrays featuring building blocks of controlled morphologies, sizes, aspect ratios and orientations at nano-, meso-, and micro-scale are genuinely fabricated without template, surfactant, undercoating, or applied external field from the hydrolysis-condensation of aqueous metal salts and complexes at mild temperatures, ca. below $100°C$[2].

In-depth investigations of the electronic structure of such novel nanostructures have been carried out by x-ray spectroscopies at synchrotron radiation facilities. Such studies include x-ray photoelectron[3] spectroscopy as well as soft x-ray absorption and emission[4] measurements, including polarization dependent[5] and energy dependent resonant inelastic x-ray scattering[6] experiments. The electron structure mapping results reveal important fundamental knowledge of orbital character and symmetry, bandgap, and quantum confinement effects of direct relevance for semiconductor technologies, optoelectronics, gas sensors, photovoltaics and photocatalysis for solar hydrogen generation.

FABRICATION METHOD

Our strategy to generate large-area of advanced nano and micro-particulate thin films at low cost and large scale is a bottom-up aqueous chemical growth approach[7] that is well-supported by a classical thermodynamic model, monitoring the nucleation, growth, and aging processes via the experimental control of the interfacial free energy of the system[8].

RESULTS AND DISCUSSION

Such a strategy has been well-illustrated on the size control of magnetite nanoparticles[9] over an order of magnitude, i.e. between 1 and 10nm. This concept and synthetic method allows the design and the creation of metal oxide nanomaterials with novel morphology, texture, and orientation which enables to probe, tune, and optimize their physical properties[10]. Particulate thin films and 3-D arrays are obtained by direct growth onto various substrates from the condensation of aqueous precursors at low temperatures. Such an approach to material synthesis offers the ability to generate anisotropic nanoparticles and to control their orientation onto substrates. Such an approach has been successfully applied for the growth of advanced nano and micro-particulate metal oxide materials such as three-dimensional arrays of ferric oxide nanorods[11] for photovoltaic[12] and photocatalytic[13] applications, nanocomposite arrays of trivalent iron and chromium corundum oxides[14] as well as oriented nanorod-[15], microrod-[16] and microtube-[17] arrays of zinc oxide for photovoltaic, optoelectronic devices and gas sensors[18]. Ferromagnetic 3-D array of metallic iron nanorods for magnetic devices[19], 2-D arrays of chromium oxide[20] for non-linear optics and magnetoelectrics applications, 3-D oriented arrays of tin dioxide nanorods with square cross-section[21] for gas sensing and photocatalytic applications have also been fabricated by such a method. An example of the variety of metal oxide structures that can be obtained with such an approach is given in figure 1.

Figure 1 Field-emission scanning electron microscopy images of crystalline oriented arrays of SnO_2 (a); ZnO (b); and α-Fe_2O_3 (c) grown onto various substrates by aqueous chemical growth

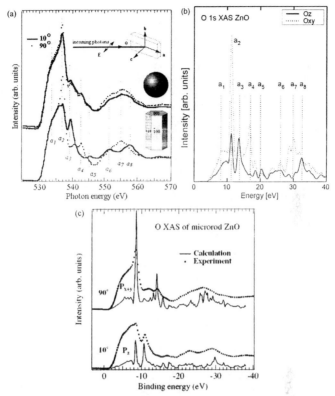

Figure 2 (a) Polarization-dependent x-ray absorption spectra of ZnO (Wurtzite structure) 3-D arrays consisting of isotropic (spherical) and anisotropic (rod) morphology. The inset illustrates the XAS experimental geometry, where *a-, b*-axes define the sample surface plane, the *c*-axis is along the growth direction of the ZnO rods, E is the polarization of incoming photons and θ indicates the angle of incidence with respect to the sample surface: 10^0 (E // *c*, lines) or 90^0 (E ⊥ *c,* dots); (b) Calculated polarization dependent O 1s XAS of wurzite bulk ZnO, where energy is referred to valence-band maximum[22]; (c) Experimental and calculated x-ray absorption spectra of ZnO rods. The calculations of the XAS spectra include the DOS and the probability of transition to the Op_{x+y} and Op_z states[5].

Synchrotron radiation studies of the electronic structure of such materials have been carried out at beamline 7.01 at the Advanced Light Source (ALS), Lawrence Berkeley National Laboratory. Such beamline is equipped with a 99-pole, 5-cm period undulator and spherical grating monochromator. For instance, the determination of orbital symmetry and orbital character of the conduction band as well as the bandgap of II-VI semiconductor ZnO (wurtzite structure) has been investigated. Strong polarization effects have been recorded on oriented nanorod arrays (fig. 2a) compared to spherical nanoparticles[5]. The measured XAS peaks are confirmed by the calculated polarization dependent O-xy and O-z 1sXAS spectra[22] of bulk ZnO (Fig. 2b, 2c). The XAS has relatively strong in-plane character at energies of around 11, 17, 25 and 29 eV, and out-of-plane character at energies of around 11, 14, 20, and 32 eV. The broad measured XAS between peak a1 and peak a2 is not seen in the calculated spectrum. It may partly be broadening and/or excitonic effects, but may also arise from native donor-like defects.

Direct assessment of the bandgap by probing the occupied (by XES) and unoccupied levels (by XAS) of ZnO nanostructures and single crystals has also been carried out[4]. The XES spectra of bulk and nanostructured ZnO are displayed together with the corresponding XAS spectrum in Fig. 3. The O K emission spectrum reflects the O $2p$ occupied states (valence band), and the O $1s$ absorption spectrum reflects the O $2p$ unoccupied states (conduction band). In the photon energy region of 530–539 eV, the x-ray absorption can be mainly assigned to the O $2p$ hybridized with Zn $4s$ states. In the region of 539–550 eV the spectrum is mainly attributed to O $2p$ hybridized with Zn $4p$ states. Above 550 eV, the contribution is mainly coming from O $2p$–Zn $4d$ mixed states. Stronger s-p-d hybridization was revealed in nanostructured ZnO since the contributions of features at 520 eV and 523 eV are enhanced. A well defined band gap can be observed between the valence-band maximum and conduction-band minimum. Our *absorption-emission* spectrum yields the fundamental band-gap energy of 3.3 eV, which is in agreement with the 3.4 eV found for bulk ZnO.

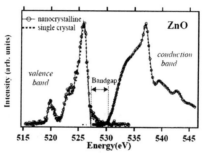

Figure 3 Oxygen x-ray *absorption-emission* spectrum reflected conduction band and valence band near the Fermi level of ZnO nanoparticles in comparison with bulk ZnO[4].

The electronic structure of quantum rod arrays of hematite has also been investigated[6]. Their Fe $2p$ absorption spectrum is displayed in figure 4a. The spectral shape appears very similar to previous XAS measurements conducted on polycrystalline or single-crystal samples. The typical spectrum shows the spin-orbit interaction of the $2p$ core level that splits the L_2 ($2p_{1/2}$) and L_3 ($2p_{3/2}$) edges, and the p-d and d-d Coulomb and exchange interactions that cause multiplets within the edges. The ligand field splitting of $3d$ transition metals, being of the same order of magnitude as p-d and d-d interactions (1-2 eV), gives a 1.4-eV-energy splitting between the t_{2g} (xy, yz, xz) and e_g (x^2-y^2, $3z^2$-r^2) orbitals. Figure 4b shows the Fe L-emission spectrum recorded with a higher photon-energy excitation (*ca* 750 eV). The spectral shape shows two peaks originating from the transitions of $3d$ orbitals to $2p_{1/2}$ and $2p_{3/2}$ core levels. A branching ratio (L_β/L_α) of 0.8 is found for the α-Fe$_2$O$_3$ bundled nanorod arrays, which appears substantially larger than that of the single crystal. The RIXS spectrum recorded at the Fe L-edge of α-Fe$_2$O$_3$

nanorods is shown in figure 4c. Several energy-loss features are clearly resolved. The low energy excitations, such as the strong d-d and charge-transfer excitations, are identified in the region from 1 to 5 eV. The 1-eV energy-loss features originate from multiple excitation transitions. The 4.1 and 6.4 eV excitation originates from charge transfer between oxygen $2p$ and iron $3d$ orbitals. The 2.5-eV excitation, which corresponds to the bandgap transition of hematite, appears significantly blue shifted compared to the reported 1.9~2.2 eV bandgap of single-crystal and polycrystalline samples as suggested above by the higher L_β/L_α branching ratio observed in the L-emission spectrum (figure 3b) of the nanorods. Such direct observation of a substantial (0.3-0.6 eV) bandgap increase is successfully attributed to a 1-D (lateral dimension) quantum confinement effect in the designed bundled ultrafine nanorods of hematite (fig. 4d). Such a finding strongly suggests that these designed quantum rod-arrays would meet the bandgap and band edge requirements for solar hydrogen[23] generation without an applied bias. Further studies aiming at developing novel low-cost visible-active semiconductor photocatalysts are currently under investigation in our laboratories.

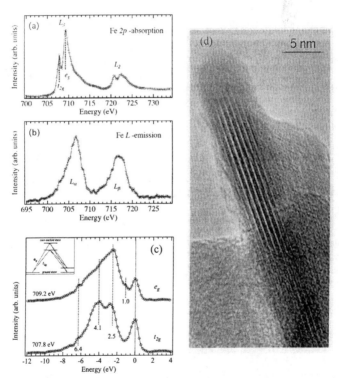

Figure 4 Fe $2p$ absorption (a), Fe L emission (b) and energy-dependent resonant inelastic x-ray scattering spectra (c) of α-Fe$_2$O$_3$ quantum rod-arrays (d). The inset (c) shows a schematic representation of the radiative de-excitation process for the two core excitations.[6]

CONCLUSION

The ability to hierarchically design and fabricate, at low cost, large three-dimensional arrays of transition metal oxides consisting of anisotropic nanoparticles with controlled orientations and aspect ratios onto various substrates without the use of templates, surfactants, or applied external fields contributes not only to the development of smart and functional nanomaterials for optoelectronic, photoelectrochemical, gas sensors and magnetic devices, but also to a better fundamental understanding of orbital character and symmetry contributions, bandgap and quantum confinement effects of one-dimensional anisotropic nanostructures by in-depth studies of their electronic structure by soft x-ray spectroscopies at synchrotron radiation facilities.

REFERENCES

[1]L. Vayssieres, "On the thermodynamic stability of metal oxide nanoparticles in aqueous solutions", *Int. J. Nanotechnol.* 2005, *2(4)*, 411-439

[2]L.Vayssieres, "On the design of advanced metal oxide nanomaterials", *Int. J. Nanotechnol.* 2004, *1(1-2)*, 1-41

[3]A. Henningsson, A. Stashans, A. Sandell, H. Rensmo, S. Södergren, L.Vayssieres, A. Hagfeldt, S. Lunell, H. Siegbahn, "Proton insertion in Polycrystalline WO_3 studied with electron spectroscopy and semi-empirical calculations", in *Advances in Quantum Chemistry*, edited by E. J. Brandas and E. Brandas (Academic Press, 2004), Vol. 47, pp. 23-36

[4]C. L. Dong, C. Persson, L. Vayssieres, A. Augustsson, T, Schmitt, M. Mattesini, R. Ahuja, C. L. Chang, J.-H. Guo "The electronic structure of nanostructured ZnO from x-ray absorption and emission spectroscopy and the local density approximation", *Phys. Rev. B* 2004, *70(19)*, 195325

[5]J.-H. Guo, L.Vayssieres, C. Persson, R. Ahuja, B. Johansson, J. Nordgren, "Polarization-dependent soft-x-ray absorption of highly oriented ZnO microrods" *J. Phys.: Condens. Matter* 2002, *14(28)*, 6969-6974

[6]L.Vayssieres, C. Sathe, S. M. Butorin, D. K. Shuh, J. Nordgren, J.-H. Guo, "One-dimensional quantum-confinement effect in α-Fe_2O_3 ultrafine nanorod arrays", *Adv. Mater.* 2005, *17(19)*, 2320-2323

[7]L.Vayssieres, "Designing ordered nano-arrays from aqueous solutions", *Pure Appl. Chem.* 2006, *78(9)* 1745-1751

[8]L. Vayssieres, "Précipitation en milieu aqueux de nanoparticules d'oxydes: Modélisation de l'interface et contrôle de la croissance", *PhD. Thesis,* Université Pierre et Marie Curie, Paris, France (1995), pp. 1-145

[9]L.Vayssieres, C. Chaneac, E. Tronc, J.P. Jolivet, "Size tailoring of magnetite particles formed by aqueous precipitation: An example of thermodynamic stability of nanometric oxide particles", *J. Colloid Interface Sci.* 1998, *205(2)*, 205-212

[10]L.Vayssieres, "Advanced semiconductor nanostructures", C. R. Chimie 2006, 9(5-6), 691-701

[11]L.Vayssieres, N. Beermann, S.-E. Lindquist, A. Hagfeldt, "Controlled aqueous chemical growth of oriented three-dimensional nanorod Arrays: Application to iron(III) oxides", *Chem. Mater.* 2001, *13(2)*, 233-235

[12]N. Beermann, L.Vayssieres, S.-E. Lindquist, A. Hagfeldt, "Photoelectrochemical studies of oriented nanorod thin films of Hematite", *J. Electrochem. Soc.* 2000, *147(7)*, 2456-2461

[13]T. Lindgren, H. Wang, N. Beermann, L.Vayssieres, A. Hagfeldt, S.-E. Lindquist, "Aqueous photoelectrochemistry of hematite nanorod-array", *Sol. Energy Mater. Sol. Cells* 2002, *71(2)*, 231-243

[14]L.Vayssieres, J.-H. Guo, J. Nordgren, "Aqueous chemical growth of αFe_2O_3-αCr_2O_3 nanocomposite thin films", *J. Nanosci. Nanotechnol.* 2001, *1(4)*, 385-388

[15]L.Vayssieres, "Growth of arrayed nanorods and nanowires of ZnO from aqueous solutions", *Adv. Mater.* 2003, *15(5)*, 464-466

[16]L.Vayssieres, K. Keis, S.-E.Lindquist, A. Hagfeldt, "Purpose-built anisotropic metal oxide material: 3D highly oriented microrod-array of ZnO", *J. Phys. Chem. B* 2001, *105(17)*, 3350-3352

[17]L.Vayssieres, K. Keis, A. Hagfeldt, S.-E. Lindquist, "Three-dimensional array of highly oriented crystalline ZnO microtubes", *Chem. Mater.* 2001, *13(12)*, 4395-4398

[18]J. X. Wang, X. W. Sun, Y. Yi, H. Huang, Y. C. Lee, O. K. Tan, and L.Vayssieres, "Hydrothermally grown ZnO nanorod arrays for gas sensing applications", *Nanotechnology* 2006, *17(19)*, 4995-4998

[19]L.Vayssieres, L. Rabenberg, A. Manthiram, "Aqueous chemical route to ferromagnetic 3D arrays of iron nanorods", *Nano Lett.* 2002, *2(12)*, 1393-1395

[20]L.Vayssieres, A. Manthiram, "2-D mesoparticulate arrays of α-Cr_2O_3", *J. Phys. Chem. B* 2003, *107(12)*, 2623-2625

[21]L.Vayssieres, M. Graetzel, "Highly ordered SnO_2 nanorod-arrays from controlled aqueous growth", *Angew. Chem. Int. Ed.* 2004, *43(28)*, 3666-3670

[22]C. Persson, C.L. Dong, L. Vayssieres, A. Augustsson, T, Schmitt, M. Mattesini, R. Ahuja, J. Nordgren, C.L. Chang, A. Ferreira da Silva, J.-H. Guo "x-ray absorption and emission spectroscopy of ZnO nanoparticles and highly oriented ZnO microrod arrays", *Microelectronics J.* 2006, *37(8)*, 686-689

[23]*Solar Hydrogen and Nanotechnology*, L.Vayssieres (Editor), Proceedings of SPIE-*The International Society for Optical Engineering*, SPIE Press Ltd., 2006, Vol. 6340, 336 pages

Author Index

Author Index